W0036736

A Working Guide to
Process Equipment

A Working Guide to Process Equipment

Editor
_____

Meenakshi Awasthi

scitus
academics

A Working Guide to Process Equipment

Edited by **Meenakshi Awasthi**

Printed in 2017

ISBN: 978-1-68117-326-9

Library of Congress Control Number: 2015939239

© 2016 by

SCITUS Academics LLC,
616, Corporate Way, Suite 2, 4766,
Valley Cottage, NY 10989

www.scitusacademics.com

This book contains information obtained from highly regarded resources. Copyright for individual articles remains with the authors as indicated. All chapters are distributed under the terms of the Creative Commons Attribution License, which permits unrestricted use, distribution, and reproduction in any medium, provided the original author and source are credited.

Notice

Reasonable efforts have been made to publish reliable data and views articulated in the chapters are those of the individual contributors, and not necessarily those of the editors or publishers. Editors or publishers are not responsible for the accuracy of the information in the published chapters or consequences of their use. The publisher believes no responsibility for any damage or grievance to the persons or property arising out of the use of any materials, instructions, methods or thoughts in the book. The editors and the publisher have attempted to trace the copyright holders of all material reproduced in this publication and apologize to copyright holders if permission has not been obtained. If any copyright holder has not been acknowledged, please write to us so we may rectify.

Contents

Preface vii

Chapter 1   Copper Recovery from Barren Cyanide Solution by Using
             Electrocoagulation Iron Process 1
             José R. Parga, Guillermo Tiburcio Munive, Jesús L. Valenzuela,
             Víctor V. Vazquez, and Gregorio González Zamarripa

Chapter 2   Large Eddy Simulation for Dispersed Bubbly Flows: A Review 21
             M. T. Dhotre, N. G. Deen, B. Niceno, Z. Khan, and J. B. Joshi

Chapter 3   Corrosion and Inhibition Effects of Mild Steel in Hydrochloric
             Acid Solutions Containing Organophosphonic Acid 81
             Manish Gupta, Jyotsna Mishra, and K. S. Pitre

Chapter 4   Modeling and Control of Distillation Column in a Petroleum
             Process 95
             Vu Trieu Minh and Ahmad Majdi Abdul Rani

Chapter 5   Quality of Electroless Ni-P (Nickel-Phosphorus) Coatings
             Applied in Oil Production Equipment with Salinity 113
             Fernando B. Mainier, Maria P. Cindra Fonseca, Sérgio S. M.
             Tavares, and Juan M. Pardal

Chapter 6   Pre-Treatment of High Free Fatty Acids Oils by Chemical
             Re-Esterification for Biodiesel Production—A Review 133
             Godlisten G. Kombe, Abraham K. Temu, Hassan M. Rajabu,
             Godwill D. Mrema, Jibrail Kansedo, and Keat Teong Lee

Chapter 7   Reactivity Investigation on Iron-Titanium Oxides for a Moving
             Bed Chemical Looping Combustion Implementation 149
             Diana C. Campos, Jamal Belkouch, Mourad Hazi, and
             Aïssa Ould-Dris

Chapter 8   The Uses of Passive Measurement of Acoustic Emissions from
             Chemical Engineering Processes 175
             Jonathan W.R. Boyd and Julie Varley

Chapter 9   **Investigation of Inlet Gas Streams Effect on the Modified Claus Reaction Furnace** 217

Reza Rezazadeh and Sima Rezvantalab

Citations..239
Index..243

Preface

A Working Guide to Process Equipment for the latest diagnostic tips, practical examples, and detailed illustrations for pinpointing trouble and correcting problems in chemical process equipment. This updated classic contains new chapters on Control Valves, Cooling Towers, Waste Heat Boilers, Catalytic Effects, Fundamental Concepts of Process Equipment, and Process Safety. Filled with worked-out calculations, the book examines everything from trays, reboilers, instruments, air coolers, and steam turbines…to fired heaters, refrigeration systems, centrifugal pumps, separators, and compressors.

Editor

Copper Recovery from Barren Cyanide Solution by Using Electrocoagulation Iron Process

José R. Parga[1], Guillermo Tiburcio Munive[2], Jesús L. Valenzuela[2], Víctor V. Vazquez[2], and Gregorio González Zamarripa[3]

[1]Institute Technology of Saltillo, Saltillo, Mexico

[2]Departament of Chemistry and Metallurgy, University of Sonora, Hermosillo, Mexico

[3]Faculty of Metallurgy, University of Coahuila, Monclova Coah, México

ABSTRACT

This paper is a brief overview of the role of inducing the nucleated electro winning of copper by using iron electrodes in electrocoagulation (EC) process. Cyanide compounds are widely used in gold ore processing plants in order to facilitate the extraction and subsequent concentration of the precious metal. Owing to cyanide solution employed in gold processing, effluents generated have high contents of free cyanide as well as copper cyanide complexes, which lend them a high degree of toxicity. In this regard, two options for the treatment of cyanide barren solutions has been used; in two ways; first for cyanide destruction by oxidation with the use of the EC process, in theory, has the advantage of decomposing cyanide at the anode and collecting copper simultaneously by a sludge of copper magnetic iron. In both cases excellent performance can be achieved using the high capacity of the bipolar iron EC technology. We found that it is possible to reduce the copper cyanide complex from 720 mg·l^{-1} **to below 10 mg·l^{-1} within 20 minutes.**

INTRODUCTION

Due to the dwindling resources of simple cyanide extractable gold deposits, a large proportion of the gold processed in the 21st century will be recovered from complex gold ores, many of which will contain soluble copper minerals. It has been estimated that about 20% of all gold deposits have significant copper mineralization commonly associated with chalcopyrite, tetrahedrite, tennantite, enargite as well as bornite and chalcocite in certain ores [1]. It has also been found that the majority of copper minerals including copper oxides, carbonates, sulfides (with the exception of chalcopyrite) and native copper are highly soluble in cyanide solutions [2]. These copper containing minerals are problematic because, when ores containing such minerals are leached with cyanide to recover the gold, copper also dissolves to form stable copper cyanide complexes. The dissolution of copper consumes a substantial quantity of cyanide and thus, if not recovered imposes a significant financial cost on the gold mine. The presence of copper also causes other problems such as competition with gold to adsorb on carbon unless a sufficient free cyanide concentration is maintained, depletion of gold electrowinning cell efficiency, and gold losses by cementation onto certain copper minerals. Ores containing greater than 0.5% reactive copper may be generally considered uneconomical to process via conventional cyanidation due to the high reagent cost. Therefore, it is necessary to reduce the amount of copper, especially in leaching circuits, in order to increases the dissolution of gold and silver, to achieve this goal it is possible to remove the copper from barren solution after Merril-Crowe process using the electrocoagulation process (EC). This process can be interesting for the copper cyanide removal from processed solutions after the Merrill Crowe process, also has low cost of operation and investment.

COPPER CYANIDE CHEMISTRY

The major challenges to the processing of gold-copper ores using cyanidation is that of the high cyanide consumptions that are typically experienced, along with effective control of the leach, particularly when there is variable cyanide-soluble copper in the ore. It is widely accepted that gold dissolution in cyanide solutions occurs as sequence of two reactions shown in Equations (1) and (2), Elsner's equation shows that oxygen is critical for the dissolution of gold.

$$2Au + 4NaCN + O_2 + 2H_2O \rightarrow$$
$$2Na\left[Au(CN)_2\right] + 2NaOH + H_2O_2$$
$$(1)$$

$$2Au + 4NaCN + H_2O_2 \rightarrow 2Na\left[Au(CN)_2\right] + 2NaOH$$
$$(2)$$

The stoichiometry of the process shows that 4 moles of cyanide are needed for each mole of oxygen present in solution. At room temperature and standard atmospheric pressure, approximately 8.2 mg of oxygen are present in one liter of water. This corresponds to 0.27×10^{-3} mol/L accordingly, the sodium cyanide concentration (molecular weight of NaCN = 49) should be equal to $4 \times 0.27 \times 10^{-3} \times 49 = 0.05$ g/L or approximately 0.01%. This was confirmed in practice at room temperature by a very dilute solution of NaCN of 0.01% - 0.5% for ores, and for concentrates rich in gold and silver of 0.5% - 5% [3]. Also, lime is added to keep the system at an alkaline pH of 10.5 - 11.0. Other factors affecting gold leaching kinetics are grain size, agitation speed, temperature, pressure, foreign ions and cyanicides.

Free cyanide exists as the uncomplexed cyanide ion, CN^-, and molecular hydrogen cyanide, HCN. These species are related by the acid dissociation of HCN:

$$HCN_{(aq)} = CN^- + H^+$$
$$(3)$$

The concentration of free cyanide is the sum of the CN⁻ and HCN concentrations, and the equilibrium diagram shown in Figure 1 illustrates the distribution.

This figure shows the proportions of free cyanide as CN⁻, and HCN as a function of pH at 25°C. At pH values below 7, cyanide is predominantly present as the unionized HCN molecule, which is easily volatilized because of its high vapor pressure. The equilibrium is displaced in favor of cyanide ion formation at pH values above 7.

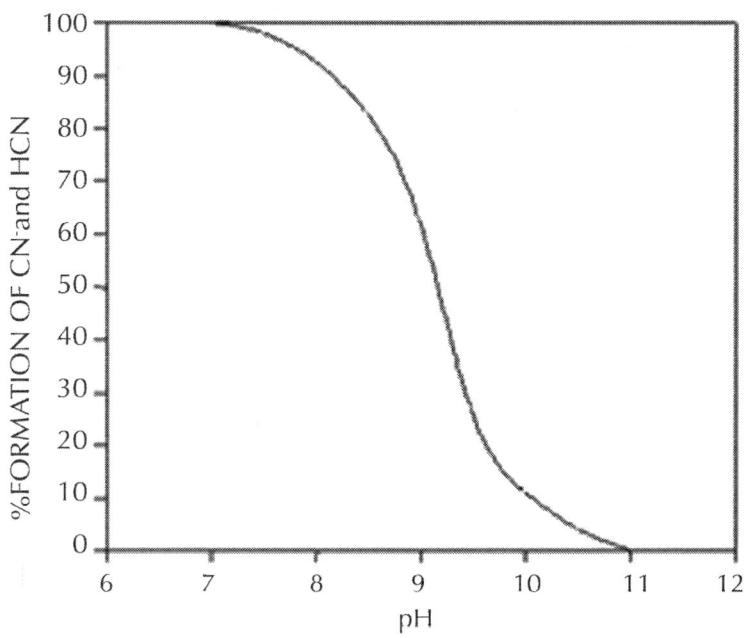

Figure 1: Equilibrium distribution diagram for cyanide as a function of pH.

Hydrogen cyanide (HCN), also known as hydrocyanic acid, is a colorless gas or liquid with a boiling point of 25.7°C, a vapor pressure of 100 kPa at 26°C and Henry's Law constant of 6.4 atm/mole [4], this makes HCN very volatile. Thus, low pH, high temperature, low pressure, and intimate contact with air, all tend to increase the rate of dissipation of cyanide from solution as hydrogen cyanide. In addition to free cyanide, other complexes such as the metal cyanide complexes formed with gold, silver, copper, nickel, iron and cobalt must be considered.

In cyanidation plants all around the world, the concentration of cyanide used to dissolve gold in ores is typically higher than the stoichiometric ratio, due to the solubility of other minerals. Free cyanide produces complexes with several metallic species, especially transition metals, which show a broad variation in both stability and solubility. Many common copper minerals are soluble in the dilute cyanide solution under typical of leach conditions found in the gold cyanidation process. Minerals such as azurite and malacite, are rapidly leached and are soluble in dilute cyanide solutions.

Enargite and chalcopyrite leach more slowly but are sufficiently soluble to cause excessive cyanide loss and contamination of the pregnant leach solutions. In reactions in aqueous solutions the cupric ion is rapidly converted to cuprous form and then copper forms a series of extremely stable soluble complexes in cyanide such as:

$$Cu^+ + CN^- = CuCN \tag{4}$$

$$CuCN + CN^- = Cu(CN)_2^- \tag{5}$$

$$Cu(CN)_2^- + CN^- = Cu(CN)_3^{2-} \tag{6}$$

$$Cu(CN)_3^{2-} + CN^- = Cu(CN)_4^{3-} \tag{7}$$

Under typical gold cyanidation conditions $Cu(CN)_3^{2-}$ has been shown to be the dominant species from the EhpH diagram for the copper-cyanide-water system [4,5]. The high consumption of cyanide during the cyanidation of copper-gold ores is due to the fact that copper forms complexes of high coordination numbers with cyanide (Reaction 3 to 6), $Cu(CN)_3^{2-}$ in particular. Therefore, hydrometallurgical treatment of these ores by cyaniding as a rule gives rise to a series of difficulties associated with increase in the cyanide consumption and decrease in the dissolution rate of gold and silver, and in the cementation process. This precipitate is of low quality, because the copper is precipitated along with gold and silver, resulting in a higher consumption of zinc dust, fluxes in the smelting of the precipitate and shorter life for crucibles.

In this regard a study is proposed to remove copper cyanide ions with, a very promising electrochemical treatment technique, which does not require chemical additions. This process is electrocoagulation (EC). The EC process operates on the principle that coagulation of copper cyanide ions from barren solutions from the MerrillCrowe process is caused by the combined effects of electrolysis gases (H_2 and O_2) and the electrolytic production of cations from the iron anodes that corrode during electrolysis.

ELECTROCOAGULATION FUNDAMENTALS

The EC process operates on the principle that the cations produced electrolytically from iron and/or aluminum anodes enhance the coagulation of contaminants from an aqueous medium. Electrophoretic motion tends to concentrate negatively charged particles in the region of the anode and positively charged ions in the region of the cathode. The consumable, or sacrificial, metal anodes are used to continuously produce polyvalent metal cations in the vicinity of the anode. These cations neutralize the negative charge of the particles carried toward the anodes by electrophoretic motion, thereby facilitating coagulation. In the flowing EC techniques, the production of polyvalent cations from the oxidation of the sacrificial anodes (Fe and Al) and the electrolysis gases (H_2 and O_2) works in combination to flocculate the coagulant materials [6], the gas bubbles produced by the electrolysis carry the pollutant to the top of the solution where it is concentrated, collected and removed. Figure 2 illustrates the schematic diagram of the process.

Generally, in the EC process bipolar electrodes are used. Pretorius [7] and Mameri [8] have reported on the use of cells with bipolar electrode arranged in series. These cells, operated at relatively low current densities, produce iron or aluminum coagulant in a more effective, fast and economical manner when compared to chemical coagulation. The savings result from a reduction in the time necessary for the treatment and an increase in the anodic area improved the efficiency of the electrolysis [7,8]. An electrocoagulation cell with bipolar electrodes connected in series is shown in Figure 3. The inner electrodes are bipolar, that is they carry both positive and negative charges on opposing faces. These charges develop on the electrode surfaces and are opposite in sign of the charge carried by the parallel electrode surface (refer to Figure 3). During electrolysis the positive side of these bipolar electrodes undergoes anodic reactions, while on the negative side, cathodic reactions occur.

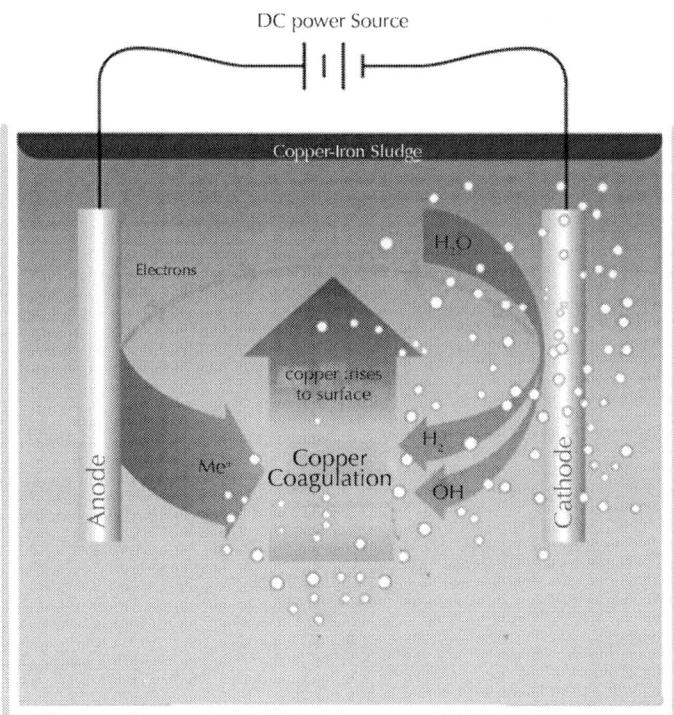

Figure 2: Schematic diagram of the EC process.

The released ions neutralize the charge of the particles and thereby initiate coagulation. The bipolar arrangement reduces the time needed for the treatment due to the increase in surface area mentioned above.

This arrangement also has the practical advantage of simplified set-up in that only two monopolar electrodes are connected to the electric power source with no interconnections between the inner bipolar electrodes.

The Chemical Reactions of the Electrocoagulation Process

The chemicals reactions that have been proposed to describe the mechanism of EC for the production of $H_{2(g)}$ and $OH^-_{(ac)}$ (cathode) and $H^+_{(ac)}$ (anode) [9] are:

$$Fe \rightarrow Fe^{3+} + 3e^-$$

(8)

$$Fe(OH)_2^+ + H_2O \leftrightarrow Fe(OH)_3 + H^+$$

(9)

$$H^+ + 2e^- \leftrightarrow H_{2(g)} \uparrow$$

(10)

$$Fe(OH)_2^+ + e^- \leftrightarrow Fe(OH)_{2(aq)}$$

(11)

$$Fe(OH)_{2(aq)} + H_2O \leftrightarrow Fe(OH)_3^- + H^+$$

(12)

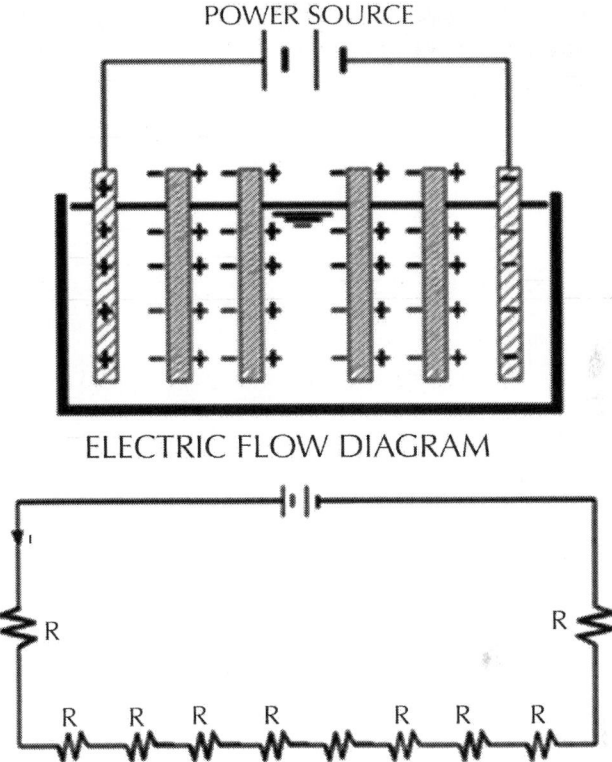

Figure 3: A schematic representation of a bipolar electrocoagulation cell and the equivalent electrical circuit (resistors reflects electrode surfaces).

$$\mathrm{Fe(OH)_3^-} \leftrightarrow \mathrm{Fe(OH)_{3(aq)}} + e^-$$

(13)

Overall reaction.

$$6\mathrm{Fe} + (12+x)\mathrm{H_2O} \leftrightarrow \frac{1}{2}(12-x)\mathrm{H_{2(g)}} \uparrow$$
$$+ x\mathrm{Fe(OH)_3} \cdot (6-X)\mathrm{Fe(OH)_{2(s)}}$$

(14)

The pH of the medium usually rises as a result of this electrochemical process and the Green Rust formed [xFe(OH)$_3$*(6-x)Fe(OH)$_{2(s)}$] remains in the aqueous stream as a gelatinous suspension, which can remove the gold and silver from pregnant cyanide rich solutions, either by complexation or by electrostatic attraction followed by coagulation and flotation.

Formation of rust (dehydrated hydroxides) occurs a while after the process, as shown in the following:

$$2Fe(OH)_3 \leftrightarrow Fe_2O_3 + 3H_2O \quad \text{(hematite)} \tag{15}$$

$$Fe(OH)_2 \leftrightarrow FeO + H_2O \tag{16}$$

$$2Fe(OH)_3 + Fe(OH)_2 \leftrightarrow Fe_3O_4 + 4H_2O \quad \text{(magnetite)} \tag{17}$$

$$Fe(OH)_3 \leftrightarrow FeO(OH) + H_2O \quad \text{(goethite, lepidocrocite)} \tag{18}$$

A schematic representation of these reactions in an EC process, using iron electrodes, is shown in Figure 4. As mentioned above, the gas bubbles produced by electrolysis carry the copper along with the sludge to the top of the solution where it is collected and removed [10]. However, it is the reactions of the metal ions that enhance the formation of the coagulant. The metal cations are hydrolyzed, releasing hydrogen ions that result in hydrogen evolution at the cathode, to yield both soluble and insoluble hydroxides that will react with or adsorb copper from the cyanide solution and also contribute to coagulation by neutralizing the negatively charged colloidal particles that may be present at neutral or alkaline pH.

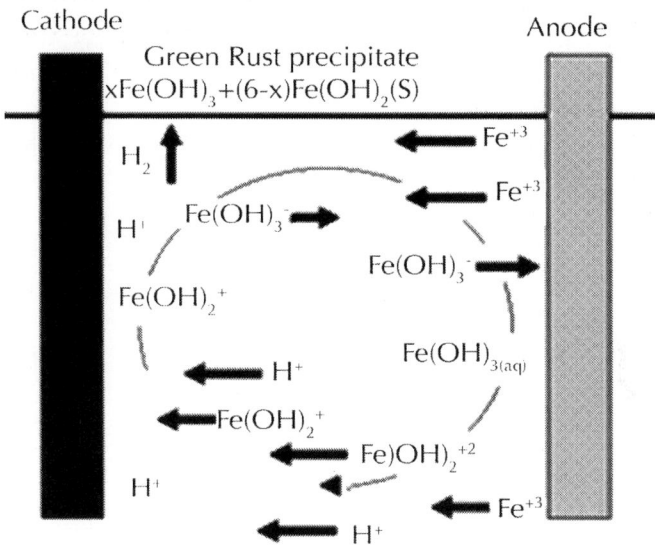

Figure 4: An illustration of the EC mechanism (arrow indicate the migration of ions, the H_2 evolution and the formation of green rust).

This enables the particles to approach closely and agglomerate under the influence of Van der Waals attracttive forces. The pH of the medium rises as a result of this electrochemical process and the Fe(OH)$_{n(s)}$ formed remains in the aqueous stream as gelatinous suspension, which can remove the $Cu(CN)_3^{2-}$ from the barren solution, either by complexation or by electrostatic attraction followed of coagulation and flotation [11]. Generally, in the EC process, bipolar electrodes are used. It has been reported that cells with bipolar electrodes, connected in series and operating at relatively low current densities [12], produce iron or aluminum coagulant more effecttively, more rapidly and more economically when compared to chemical coagulation.

EXPERIMENTAL DETAILS

The experimental work was performed using a barren solution from the Merrill Crowe process containing an average in the range of 660 - 712 ppm of copper. EC experiments were performed using a 600 ml Pyrex beaker glass (Figure 5), equipped with two iron electrodes (10 cm × 2.5 cm) with an electrode separation of 5 mm. A regulated power supply (model Steren PRL-25) to provide the necessary energy to the electrocoagulation cell, magnetic stirrer (Model PC-310, Corning) with 100 rpm was used, and the initial and final pH was taken with a pH meter (VWR Scientific 8005), filtration was performed with Whatman filter paper No.42. To determine the adsorption copper in iron hydroxide species generated at the anode, a barren solution after the cementation process was used that was provided by the William Mine Co. This solution contained an average of 0.02 mg/L gold and 0.1 mg/L silver along with amounts of zinc, lead. Among others, the analysis was performed according to EPA 200.7 (EPA 200.7 is an analytical method for detection of metals and trace elements by ICP/atomic emission spectrometry). The solution and the solids were separated by filtration through filter paper, the sludge from the EC was dried for 8 hrs in an oven at 80°C. The tables below show the initial conditions of the EC test to remove copper.

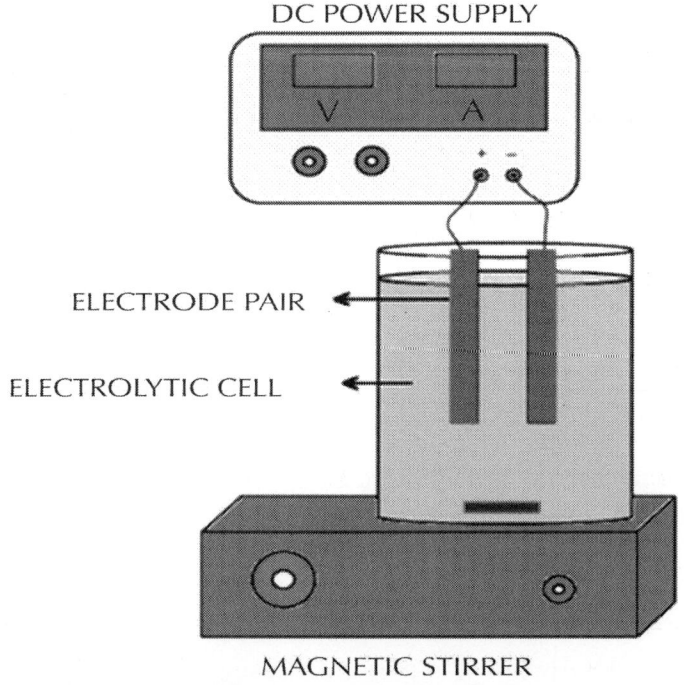

Figure 5: EC schematic diagram of the experimental setup.

Optimization of Parameters

In order to find the optimum parameters of the EC process for the removal of copper, experiments were carried out by changing the pH of the solution, residence time in the EC cell and voltage and amperage.

RESULTS

Tables 1-3 show the results for tests performed with 4 grams/liter of sodium chloride.

From these results it was determined that the optimum parameters were: pH 8, residence time of 20 minutes and 4 grams/liter of NaCl, this achieved 99% copper removal. Also, when the time increased from 15 to 20 minutes the removal of copper increased from 92% to 99%, this occurs in the pH range from 8 to 9 approximately, this coincides with the production of the magnetic iron, Fe_3O_4, which has magnetic properties that accelerates the process of adsorption of metals, the adsorption rate is then physically, because it is caused by the magnetic forces of the magnetite into copper, without altering their chemical composition. This removal of copper also can explain with a decrease in the zeta potential on iron hydroxides which causes a decrease in repulsive forces between the particles, generated collision between particles thus favors the formation of flocs which float to the water surface through micro bubbles generated from oxygen and hy drogen from the iron electrodes. Also, the advantages of the EC process is the decomposing of cyanide at the anode, where the anodic oxidation of cyanide is proportional to the alkalinity of the electrolyte and consistent with the following mechanism:

Table 1. Results obtained for pH 8 and 4 g/liter of NaCl

fInitial Conditions of EC Test					Sludge Chemical Essay		
Time (min)	Voltage (volt)	Current (A)	%Cu	%Fe	[Cu] inintial (ppm)	[Cu]final (ppm)	%Examination
5	12.1	10.3	4.6	21.3	660	328	50
10	12.3	10.5	6.2	23	664	188	72

| 15 | 11.5 | 10.9 | 9.4 | 24 | 712 | 70 | 90 |
| 20 | 12.4 | 10.1 | 13 | 27 | 716 | 5 | 99 |

Table 2: Results obtained for pH 9 and 4 g/liter of NaCl

fInitial Conditions of EC Test					Sludge Chemical Essay		
Time (min)	Voltage (volt)	Current (A)	% Cu	% Fe	[Cu] inintial (ppm)	[Cu] final (ppm)	% Examination
5	12.1	10.3	4.4	22	673	428	36
10	12.3	10.5	7	23.5	677	190	76
15	11.5	10.9	10	24	679	80	88
20	12.4	10.1	12.8	27.5	701	10	98

Table 3: Results obtained for pH 10 and 4 g/liter of NaCl

fInitial Conditions of EC Test					Sludge Chemical Essay		
Time (min)	Voltage (volt)	Current (A)	% Cu	% Fe	[Cu] inintial (ppm)	[Cu] final (ppm)	% Examination
5	12.1	10.3	4.8	21.5	667	424	36
10	12.3	10.5	6.8	22	686	186	73
15	11.5	10.9	10.8	24	680	92	86
20	12.4	10.1	13	28	705	15	98

$$CN^- + 2OH \rightarrow CNO^- + H_2O + 2e^-$$

(18)

$$2CNO^- + 4OH^- \rightarrow 2CO_2 + N_2 + 2H_2O + 6e^-$$

(19)

$$CNO^- + 2H_2O \rightarrow NH_4 + CO_3^-$$

(20)

Figure 6 shows the graph of timeelimination where it is observed the behavior of the removal of copper.

PRODUCT CHARACTERIZATION

In order to identify the iron species present, Scanning Electron Microscope (SEM/EDX) was used to characterize the solid products formed during the EC process for removal of copper with iron electrodes.

Scanning Electron Microscopy (SEM/EDAX)

Figure 7 shows SEM image and Figure 8 shows EDAX of copper adsorbed on iron species. These SEM and EDAX results show that the surfaces of these iron oxide/ oxyhydroxide particles were coated with a layer of copper.

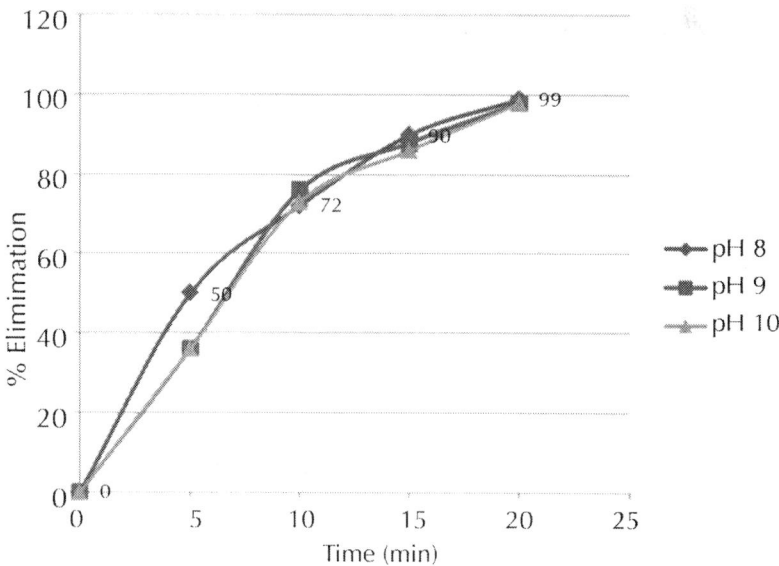

Figure 6: % Elimination of copper at pH 8, 9 and 10.

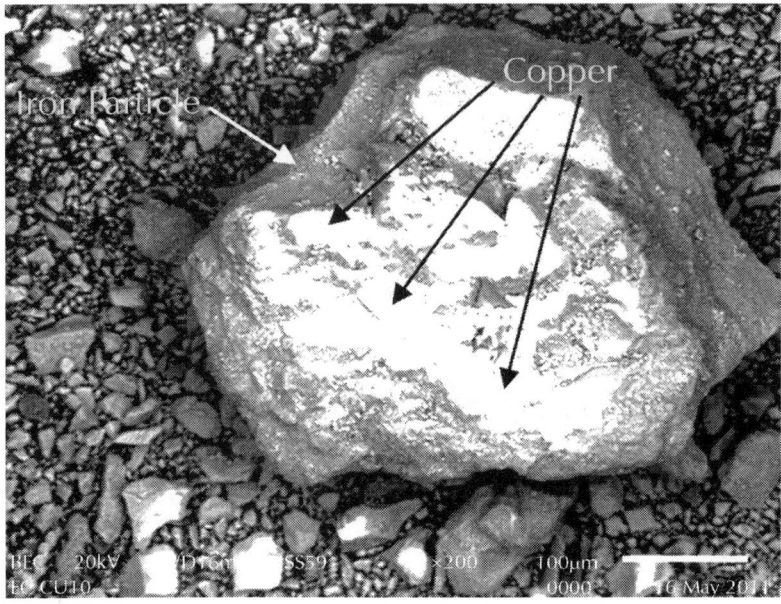

Figure 7: SEM image.

X-Ray Diffraction

Diffractograms were obtained with a Bruker AXS D4 Endeavor diffractometer operating with a Cu-K radiation source filtered with a graphite monochromator ($\lambda = 1.5406$ Å). The samples were wet ground to a fine powder (isopropyl alcohol from Sigma-Aldrich) and pressed into a sample holder. The XRD scans were recorded from $20°$ to $80°$ 2θ, with $0.02°$ step-width and with a 10 s counting time for every step-width (increment). Experiments were run at 40 kV and 40 mA power. Figure 9, shows a diffractogram of the filtered solid products (the feed solution contained 700 ppm of copper and the pH of the solution after EC was ~10).

This figure indicates the presence of magnetite, geothite, iron hydroxide oxide, and lepidocrocite in the solid products.

CONCLUSIONS

Electrocoagulation process was carried out to obtain a 99% of copper removal. The optimal operations conditions were: pH = 8, residence time: 20 minutes and 4 g/l of sodium chloride as conductivity modifier. The solid product obtained from EC process was 13% copper and 24% iron. The EC process does not generate any smell in the process of elimination of copper from the barren solution. During the EC process is not necessary to add any reagent (except Nacl), since the coagulating agent is generated in situ.

We found that it is possible to reduce the copper cyanide complex from 720 mg·l^{-1} to below 10 mg·l^{-1} within 20 minutes.

Also, the results of this study suggest that EC produces magnetic particles of magnetite. lepidocrocita and amorphous iron oxyhydroxide species that can be used to removal copper. The Scanning Electronic Microscopy, techniques demonstrate that the formed species are of magnetic type, like lepidocrocite and magnetite which adsorbed the copper particles on his surface due to the electrostatic attraction between both metals.

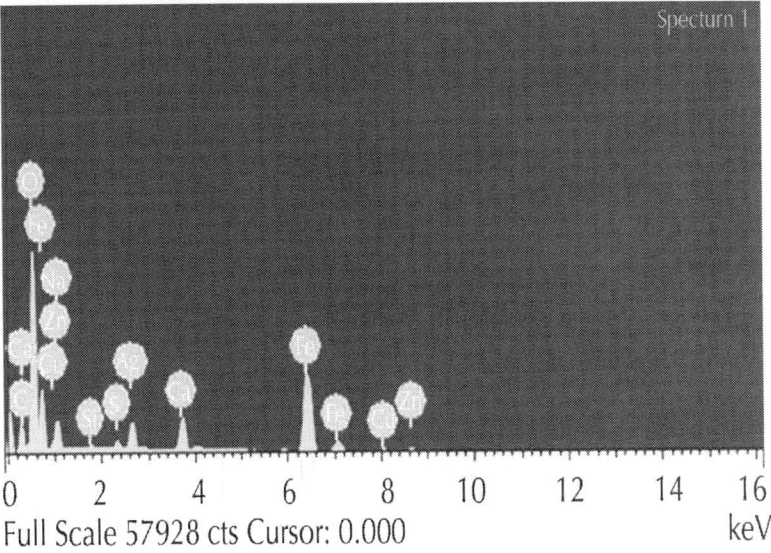

Figure 8: EDAX Analysis.

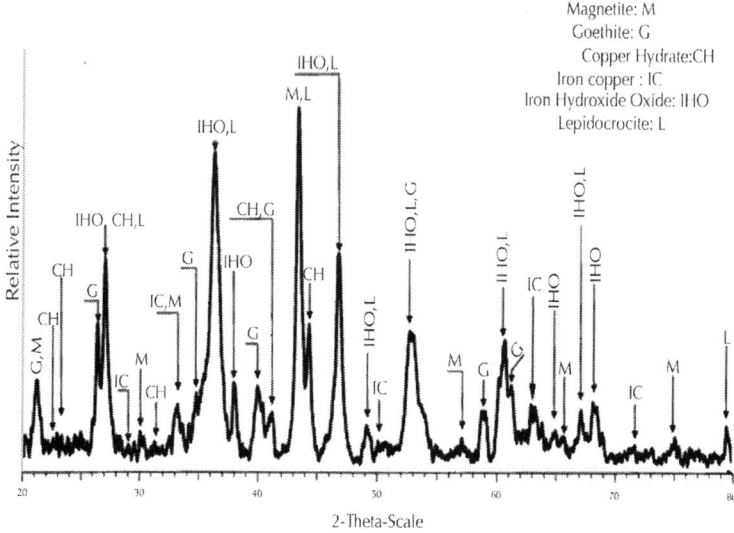

Figure 9: X-ray diffraction spectrum obtained from the EC product at pH = 10.

ACKNOWLEDGEMENTS

The authors wish to acknowledge support of this project to the National Council of Science and Technology (CONACYT) and to Dirección General de Educación Superior Tecnológica (DGEST).

REFERENCES

1.   D. M. Muir, S. R. LaBrooy and C. Cao, "Recovery of Gold Fromcopper-Bearing Ores," In: R. J. Harden, Ed., Gold Forum Ontechnology and Practic, World Gold, 1989.

2.   J. O. Marsden abd C. I. House, "The Chemistry of Gold Extraction," 2nd Edition, Society for Mining, Metallurgy and Exploration, Inc., Littleton, 2006.

3.   J. R. Parga, S. S. Shukla and D. L. Cocke, "Photocatalytic Detoxification of Cyanide and Recovery of Titanium Dioxide by Electrocoagulation," Research Journal of Chemistry and Environment, Vol. 9, No. 1, 2005, pp. 60-63.

4. S. T. Mudder, "The Chemistry and Treatment of Cyanidation Wastes," Mining Journal Books Limited, London, 1991, pp. 277-278.

5. M. D. Adams and R. Lawrance, "Biogenic Sulphide for Cyanide Recycle and Copper Recovery in Gold-Copper Ore Processing," International Workshop on Process Hydrometallurgy, Hydroprocess Brisbone, 2008, pp. 14-16.

6. O. Asare, K. Xue and T. Ciminelli, "Solution Chemistry of Cyanide Leaching Systems. In Precious Metals Mineralogy," Extraction and Processing-Proceedings of and International Symposium, 1984, pp. 173-197.

7. N. Mameri, A. R. Yeddou, H. Lounici, D. Belhocine, H. Grib and B. Bariou, "Defluoridation of Septentrional Sahara Water of North Africa by Electrocoagulation Process Using Bipolar Aluminum Electrodes," Water Research, Vol. 32, No. 5, 1998, pp. 1604-1612.doi:10.1016/S0043-1354(97)00357-6

8. W. A. Pretorius, W. G. Johannes and G. G. Lempert, "Electrolytic Iron Flocculant Production with a Bipolar Electrode in Series Arrangement," Water South Africa, Vol. 17, No. 2, 1991, pp. 133-138.

9. G. Pavas, M. P. Camargo, C. Jones and V. T. Pineda, "Oxidación Fotocatalítica de Cianuro," Universidad EAFIT, Medellín, 2005, pp. 56-58.

10. J. R. Parga, H. M. Casillas, V. Vazquez and J. L. Valenzuela, "Cyanide Detoxification of Mining Wastewaters with TiO_2 Nanoparticles and Its Recovery by Electrocoagulation," Chemical Engineering and Technology, Vol. 32, No. 12, 2009, pp. 1901-1908.doi:10.1002/ceat.200900177

11. G. Vicuña and I. Tuñon, "Apuntes de Química Avanzada," Departamento de Química-Física. Universidad de Valencia, 2006, pp. 103-115.

12. A. G. Gupta and S. Kundu, "Adsorptive Removal of As(III) from Aqueous Solution Using Iron Oxide Coated Cement (IOCC): Evaluation of Kinetic Equilibrium and Thermodynamic Models," Separation and Purification Technology, Vol. 51, No. 2, 2006, pp. 165-172.doi:10.1016/j.seppur.2006.01.007

Large Eddy Simulation for Dispersed Bubbly Flows: A Review

M. T. Dhotre[1], N. G. Deen[2], B. Niceno[3], Z. Khan[4],
and J. B. Joshi[4, 5]

[1]ABB Switzerland Ltd., 5400 Baden, Switzerland

[2]Multiphase Reactors Group, Department of Chemical Engineering and Chemistry, Eindhoven University of Technology, The Netherlands

[3]Laboratory for Thermal-Hydraulics, Nuclear Energy and Safety Department, Paul Scherrer Institute, Switzerland

[4]Institute of Chemical Technology, Matunga, Mumbai 400 019, India

[5]Homi Bhabha National Institute, Anushakti Nagar, Mumbai 400 094, India

ABSTRACT

Large eddy simulations (LES) of dispersed gas-liquid flows for the prediction of flow patterns and its applications have been reviewed. The published literature in the last ten years has been analysed on a coherent basis, and the present status has been brought out for the LES Euler-Euler and Euler-Lagrange approaches. Finally, recommendations for the use of LES in dispersed gas liquid flows have been made.

INTRODUCTION

Gas-liquid flows are often encountered in the chemical process industry, but also numerous examples can be found in petroleum, pharmaceutical, agricultural, biochemical, food, electronic, and power-generation industries. The modelling of gas-liquid flows and their dynamics has become increasingly important in these areas, in order to predict flow behaviour with greater accuracy and reliability. There are two main flow regimes in gas-liquid flows: separated (e.g., annular flow in vertical pipes, stratified flow in horizontal pipes) and dispersed flow (e.g., droplets or bubbles in liquid). In this work, we consider only dispersed bubbly flows.

Dispersed Bubbly Flow: The description of bubbly flows involves modelling of a deformable (gas-liquid) interface separating the phases; discontinuities of properties across the phase interface; the exchange between the phase; and turbulence modelling.Most of the dispersed flow models are based on the concept of a domain in the static (Eulerian) reference frame for description of the continuous phase, with addition of a reference frame for the description of the dispersed phase. The dispersed phase may be described in the same static reference frame as the continuous, leading to the Eulerian-Eulerian (E-E) approach or in a dynamic (Lagrangian) reference frame, leading to the Eulerian-Lagrangian (E-L) approach.

In the E-L approach, the continuous liquid phase is modelled using an Eulerian approach and the dispersed gas phase is treated in a Lagrangian way; that is, the individual bubbles in the system are tracked by solving Newton's second law, while accounting for the forces acting on the bubbles. An advantage here is the possibility to model each individual bubble, also incorporating bubble coalescence and breakup directly. Since each bubble path can be calculated accurately within the control volume, no numerical diffusion is introduced into the dispersed phase computation. However, a disadvantage is, the larger the system gets the more equations need to be solved, that is, one for every bubble.

The E-E approach describes both phases as two continuous fluids, each occupying the entire domain, and interpenetrating each other. The conservation equations are solved for each phase together with interphase exchange terms. The E-E approach can suffer from numerical diffusion. However, with the aid of higher order discretization schemes, the numerical diffusion can be reduced sufficiently and can offer the same order of accuracy as with E-L approach (Sokolichin et al. [1]). The advantage here is that the computational demands are far lower compared to the E-L approach, particularly for systems with higher dispersed void fractions. We review these approaches here with respect to the turbulence descriptions.

Turbulence Modelling: The major difficulty in modelling multiphase turbulence is the wide range of length and time scales on which turbulent mixing occurs. The largest eddies are typically comparable in size to the characteristic length of the mean flow. The smallest scales are responsible for the dissipation of turbulence kinetic energy. The Direct Numerical Simulation (DNS) approach, with no modelling, resolves all the scales present in turbulence. However, it is not feasible for practical engineering problems involving high Reynolds number flows. The Reynolds-Averaged Navier–Strokes (RANS) approach is more feasible; it models the time-averaged velocity field either by using turbulent viscosity or by modelling the Reynolds stresses directly.

The large eddy simulation (LES) falls between DNS and RANS in terms of the fraction of the resolved scales. In LES, large eddies are resolved directly, that is, on a numerical grid, while small, unresolved eddies are modelled. The principle behind LES is justified by the fact that the larger eddies, because of their size and strength, carry most of the flow energy (typically 90%) while being responsible for most of the transport, and therefore they should be simulated precisely (i.e., resolved). On the other hand, the small eddies have relatively little influence on the mean flow and thus can be approximated (i.e., modelled). This approach to turbulence modelling also allows a significant decrease in the computational cost over direct simulation and captures more dynamics than a simple RANS model.

In RANS models often the assumption of isotropic turbulence is made for the core of the flow, which is not valid in dispersed bubbly flows; that is, the velocity fluctuations in the gravity direction are typically twice those in the other directions. This assumption is not made in LES for large structures of the flow, giving LES an advantage over RANS for the core regions of the flow. However, the situation is different close to the walls, where LES' assumption of isotropic turbulence is heavily violated, due to the absence of large eddies close to the walls.

LES FOR DISPERSED BUBBLY FLOWS

In dispersed bubbly flows, the large-scale turbulent structures interact with bubbles and are responsible for the macroscopic bubble motion, whereas small-scale turbulent structures only affect small-scale bubble oscillations. Since, large scales (carrying most of the energy) are explicitly captured in LES and the less energetic small scales are modelled using a subgrid-scale (SGS) model, LES can reasonably reproduce the statistics of the bubble-induced velocity fluctuations in the liquid.

There are three important considerations for modelling of dispersed bubbly flows.

- Separation of length scales of the interface, that is, micro-, meso-, and macroscales. The separation of these scales forms the basis for "filtering" the Navier–Stokes equations and applying proper model equations for multiphase situation. Important for dispersed flow is to identify the scales at which the governing equations are to be applied; microscales, that is, scales which are small enough to describe individual bubble shapes; mesoscales, which are comparable to bubble sizes; and macroscales, which entail enough bubbles for statistical representation.

- The grid-scale equations. Depending on the ratio of the length scales introduced above, with the grid resolution we can afford, on a given computer hardware, a proper form of the governing equations must be chosen. For instance, if the mesh size is in the micro-scale order, one can use single-fluid, interface tracking techniques to solve the problem. If, on the other hand, the grid size is large enough for statistical description of bubbles, the E-E approach can be used. Should the grid size be comparable to the meso-scales, we are in a limiting area for both approaches, and special care must be taken in order to solve equations which describe the underlying physics consistently.

- The physical models. Depending on the selected grid-scale equations, physical models of various complexities must be employed. The options here are numerous, whether they concern turbulence modelling or interphase modelling, but these models are generally simpler in case more of the microscales are resolved.

In the following sections, we describe each of these three elements to model turbulent dispersed bubbly flow.

Filtering Operation

The aim of filtering the Navier-Stokes equations is to separate the resolved scales from the SGS (nonresolved). The interface between the phases, and the level of detail required in its resolution/modelling, defines the filter in a multiphase flow.

When LES is applied at a micro-scale, filtering of turbulent fluctuations needs to be combined with interface tracking methods. These methods have been developed and used in both dispersed flow and free surface flow by Bois et al. [2], Toutant et al. [3, 4], Magdeleine et al. [5], Lakehal [6], and Lakehal et al. [7]. These methods require that all phenomena having an influence on space and time position of the interface are also simulated. For the amount of details required and the large size of practical problems of interest, these types of models should merely be seen as a support for the modelling and validation of more macroscopic approaches and cannot address a real industrial-scale problem (Bestion [8]).

When LES is applied at a macro-scale, the interface resolution is not considered. However, in practical simulations, these would require too coarse grids, leading to poor resolution of turbulence quantities. Much more often we are in the meso-scale region, in which the mesh size is comparable to bubble sizes. This pushes the main assumptions of the E-E approach to its limit of validity, and the grid is not fine enough for full interface tracking. In other words, the mesh requirement for E-E multiphase modelling conflicts with the requirements by LES approaches [9].

The issue of the requirement of the mesh size was first addressed by Milelli et al. [11] who carried out a systematic analysis and performed a parametric study with different mesh sizes and bubble diameters. They showed that for case of a shear layer laden with bubbles it was possible to provide an optimum filter width $1.2 < \Delta/d_b < 1.5$, where Δ is the filter width and d_b is the bubble diameter (shown in Figure 1). This means that the grid space should be at least 50% larger than the bubble diameter. The constraint imposed on the ratio Δ/d_b implies that the interaction of bubbles with the smallest resolved scales is captured without additional approximation.

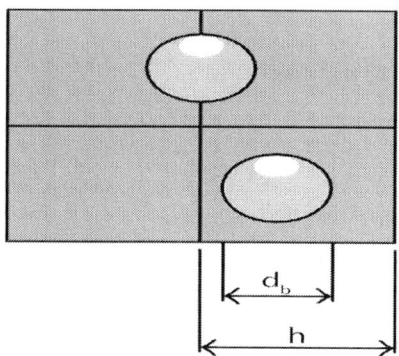

Figure 1: Milelli condition (from Niceno et al. [10]).

Grid-Scale Equations

The principle of the LES formulation is to decompose the instantaneous flow field into large-scale and small-scale components via a filtering operation. If $\overline{\phi}_f$ denotes the filtered or grid-scale component of the variable ϕ_f that represents the large-scale motion then

$$\phi_f = \underbrace{\overline{\phi}_f}_{\text{resolved}} + \underbrace{\phi'_f}_{\text{subgrid}},$$

$$(1)$$

where ϕ is the variable of interest, subscript f refers either to the liquid or the gas phase. In the remainder of this paper, we omit the bars of all resolved variables for the sake of simplicity. The following filtered equations are obtained:

$$\frac{\partial}{\partial t}\left(\alpha_f \rho_f \mathbf{u}_f\right) + \Delta \cdot \left(\alpha_f \rho_f \mathbf{u}_f\right) = 0,$$

(2)

$$\frac{\partial}{\partial t}\left(\alpha_f \rho_f \mathbf{u}_f\right) + \Delta \cdot \left(\alpha_f \rho_f \mathbf{u}_f \mathbf{u}_f\right)$$

$$= -\nabla \cdot \left(\alpha_f \tau_f\right) - \alpha_f \nabla p + \alpha_f \rho_f g + \mathbf{M}_f.$$

(3)

The right hand side terms of (3) are, respectively, the stress, the pressure gradient, gravity, and the momentum exchange between the phases due to interface forces.

The SGS stress tensor which reflects the effect of the unresolved scales on the resolved scales is modelled as

$$\tau_f = -\mu_{\text{eff},f}\left(\nabla \mathbf{u}_f + \left(\nabla \mathbf{u}_f\right)^T - \frac{2}{3}I\left(\nabla \cdot \mathbf{u}_f\right)\right),$$

(4)

where $\mu_{\text{eff},f}$ is the effective viscosity.

In the E-E approach, separate equations are required for each phase (see (3), f=l, g), together with interphase exchange terms (for details, Drew [12]). In most of the investigations, turbulence is taken into consideration for the continuous phase by SGS models. The dispersed gas phase is modelled as laminar, but influence of the turbulence in the continuous phase is considered by a bubble-induced turbulence (BIT) model.

In the E-L approach, there are two coupled parts: a part dealing with the liquid phase motion and a part describing the bubbles motion. The dynamics of the liquid are described in a similar way as in the E-E approach, whereas the bubble motion is modelled through the second law of Newton.

Since, the governing equations for the liquid and gas phase are expressed in the Eulerian and Lagrangian reference frames, respectively; a mapping technique is used to exchange interphase coupling quantities. Depending upon the volume fraction of the dispersed phase, one-way (e.g. $\alpha_g < 10^{-6}$,) or two-way coupling between gas phase to liquid phase ($10^{-6} < \alpha_g < 10^{-3}$) prevails. In both cases, bubble-bubble interactions (i.e., collisions) can be neglected, but the effect of the bubbles on the turbulence structure in the continuous phase has to be considered for higher volume fraction and does not play any role in lower volume fraction of gas phase Elgobashi [13]. The work reviewed here considers the two-way coupling which consists of the following.

Forward Coupling (Liquid to Bubble)

In the forward coupling, calculated liquid velocities, velocity gradients, and pressure gradients on an Eulerian grid are interpolated to discrete bubble locations for solving the Lagrangian bubble equation motion.

Backward or Reversed Coupling (Bubble to Liquid)

The forces available at each bubble's centroid need to be mapped back to the Eulerian grid nodes in order to evaluate the reaction force F. The two-way interaction (forward and backward) is accomplished with a mapping method, for example, PSI-cell method [14], modified PSI-wall-method [15], or mapping functions discussed by Deen et al. [16].

Interfacial Forces

The motion of a single bubble with constant mass can be written according to Newton's second law:

$$ m_b \frac{d\mathbf{v}}{dt} = \sum \mathbf{F}. $$

(5)

The bubble dynamics are described by incorporating all relevant forces acting on a bubble rising in a liquid. It is assumed that the total force, $\sum F$, is composed of separate and uncoupled contributions originating from pressure, gravity, drag, lift, virtual mass, wall lubrication and wall deformation turbulent dispersion:

$$ \sum \mathbf{F} = \mathbf{F}_P + \mathbf{F}_G + \mathbf{F}_D + \mathbf{F}_L + \mathbf{F}_{VM} + \mathbf{F}_{TD} + \mathbf{F}_{WL} + \mathbf{F}_{WD}. $$

(6)

For each force the analytical expression or a semiempirical model is used, based on bubble behaviour observed in experiment or in DNS.

To summarize, the influence/contribution of these forces are as follows.

- The modeling of the lift force for capturing bubble plume meandering and bubble dispersion is important. However there is an uncertainty regarding appropriate value or correlation representing lift coefficient. There is also recommendation that bubble size-dependent lift coefficient should be chosen [17].

- The value of the lift coefficient can be different than the one used in RANS approach. It is because of different handling of factors responsible for bubble dispersion, that is, the interaction between the bubbles and influence of turbulent eddies in the liquid phase. In RANS approach, they are considered by means of the lift and turbulent dispersion force, with uncertainty of exact contribution of the individual forces. Most of the investigators use a constant value of the lift coefficient (C_L=0.5), while the value of the turbulent dispersion coefficient is varied (0.1 to 1.0) to get good agreement with the experimental data. However, in LES, bubble dispersion caused by liquid phase turbulent eddies is implicitly calculated, and a more realistic contribution of the lift force can be used. The coefficient for the effective lift force thus may vary between the two approaches [18].

- The virtual mass force is proportional to the relative acceleration between the phases and is negligible once a pseudosteady state is reached. It has little influence on the simulation results for bubble plumes [19], Milelli [20]. It is mainly because of the acceleration and deceleration effects are restricted to small end regions of the column. A constant coefficient is used in almost all investigations.(4)In LES, through filtering, velocities are decomposed into a resolved and a SGS part. The resolved part of the turbulent dispersion is implicitly computed. However, in case of a bubble size smaller than the filter size, turbulent transport can be present at SGS level and should be considered [9]. This can be done using a one-equation model, wherein it can be modelled by replacing the total kinetic energy by SGS contribution (K_{SGS}). By the same argument, other forces also need modelling at SGS level.

The values or expressions for the coefficient of drag, lift and virtual mass force used by different investigators are given in Tables 1 and 3.

Table 1: Comparison of LES simulations

No.	Author	Column D (× W) × H (m)	Sparger design	Bubble diameter	Range of V_g, m/s	Number of grid cells	Filter	SGS Model	BIF closure models⁺	Interfacial force coefficient closures⁺⁺		
										Drag C_D	Lift C_L	Virtual mass C_{VM}
(1)	Deen et al.[19]	0.15 × 0.15 × 0.45	Perforated plate	4 mm	4.9 × 10⁻³	15 × 15 × 45 32 × 32 × 90	Δ = 5-40 mm	Smagorinsky, C_s = 0.1	(1)	(4)	0.5	0.5
(2)	Bove et al.[29]	0.15 × 0.15 × 0.45	Perforated plate	4 mm	4.9 × 10⁻³	15 × 15 × 45 9 × 6 × 30	Δ = 10-17 mm	Smagorinsky, C_s = 0.05-0.2		(2, 4)	0.5	0.5
(3)	Zhang et al.[26]	0.15 × 0.15 × 0.45-0.90	Perforated plate	4 mm	4.9 × 10⁻³	15 × 15 × 45 15 × 15 × 90		Smagorinsky, C_s = 0.1	(1, 2, 3)	(2)°	*	*
(4)	Tabib et al.[17]	0.15 × 1.0	Perforated plate	5 mm	20.9 × 10⁻³	150000	Δ = 3 mm	Smagorinsky, C_s = 0.1	(1)	(1)	⁺	*
(5)	Dhotre et al.[20]	0.15 × 0.15 × 0.45	Perforated plate	4 mm	4.9 × 10⁻³	13 × 13 × 50 15 × 15 × 100	$\bar{d}_b$ = Δ	Smagorinsky, C_s = 0.1 Germano	(1)	(1)	0.5	0.5
(6)	Niceno et al.[9]	0.15 × 0.15 × 0.45	Perforated plate	4 mm	4.9 × 10⁻³	15 × 15 × 45 30 × 30 × 90	Δ is equal to the grid spacing Δ/d_b = 1.5	Smagorinsky, C_s = 0.12 DSM	(1, 2)	(1, 5)	0.5	0.5
(7)	Dhotre et al.[18]	2.0 × 3.4	Perforated plate	2.6 mm	4.9 × 10⁻³		2.8 mm = Δ = 4.6 mm	Smagorinsky, C_s = 0.12	(1)	(5)	0.5	0.5
(8)	Niceno et al.[10]	0.15 × 0.15 × 1.0	Perforated plate	4 mm	4.9 × 10⁻³	30 × 30 × 100	Δ = 5 mm	DSM Germano	(1, 2)	(4, 5)	0.5	0.5
(9)	Tabib and Schwarz[30]	0.15 × 1.0	Perforated plate	3-5 mm	20.9 × 10⁻³			DSM	(2)	Max [(1),(6)]	0.05	
(10)	van den Hengel et al.[31]	0.15 × 1.0	Perforated plate	3 mm	4.9 × 10⁻³	15 × 15 × 45	Δ = 10 mm	Smagorinsky	(1)	(3)	0.5	0.5
(11)	Hu and Celik[32]	0.15 × 0.08 × 1.0	Pipe sparger	1.6 mm	0.660 × 10⁻³	96 × 36 × 8 130 × 80 × 10	PSI-ball method 2 mm	Smagorinsky, C_s = 0.032	(1)	(3)	0.5	0.5
(12)	Lain[27]	0.14 × 1.0	Porous membrane	2.6 mm	0.272 × 10⁻³	30 × 30 × 10 45 × 45 × 10	Δ/d_b = 1.8	Smagorinsky, C_s = 0.1	(1)	*	*	0.5
(13)	Darmana et al.[33]	0.2 × 0.03 × 1.0	Multipoint gas injection	4 mm	7.0 × 10⁻³	80 × 12 × 400	Δ = 2.5 mm	Vreman, C_s = 0.1		(3)	*	0.5
(14)	Sungkorn et al.[34]	0.15 × 0.15 × 0.45	Perforated plate	4 mm	4.9 × 10⁻³	30 × 30 × 90	Δ is equal to the grid spacing Δ/d_b = 1.25	Smagorinsky, C_s = 0.10-0.12	(6)	(1)	*	0.5
(15)	Bai et al.[35], Bai et al.[36]	0.15 × 0.15 × 0.45	Perforated plate	5 mm	5.0 × 10⁻³ 10.0 × 10⁻³ 15.0 × 10⁻³ 23.9 × 10⁻³	15 × 15 × 45 20 × 20 × 60 30 × 30 × 90		Vreman Smagorinsky C_s = 0.1		(3)	0.5	0.5

#The authors have studied the effect of this force over a range.

+Numbers indicated are refered to Table 2.

++Numbers indicated are refered to Table 3.

SGS Models

It is well known that in turbulent flow energy generally cascades from large to small scales. The primary task of the SGS model therefore is to ensure that the energy drain in the LES is same as obtained with the cascade fully resolved as one would have in a DNS. The cascading, however, is an average process. Locally and instantaneously the transfer of energy can be much larger or much smaller than the average and can also occur in the opposite direction ("backscatter").

Smagorinsky [21] Model

The simplest, well-known, and mostly used Smagorinsky [21] model is based on the Boussinesq hypothesis. It requires the definition of time and length scales and a model constant. Smagorinsky used the following expression to calculate the turbulent viscosity, that is, the SGS viscosity:

$$\mu_{\text{eff},l} = \mu_{\text{lam},l} + \rho_l (C_S \Delta)^2 \sqrt{S^2},$$

(7)

where $\mu_{\text{lam},l}$ is the (laminar) dynamic viscosity, C_S is the Smagorinsky constant, S is the characteristic strain tensor of filtered velocity, and Δ is the filter width, usually taken as the cubic root of the cell volume.

In the single-phase flow literature, the value of the constant used is in the range from $C_S = 0.065$ (Moin and Kim [22]) to $C_S = 0.25$ (Jones and Wille [23]). The value of C_S used in gas-liquid flows varies from that of single phase flow and is in the range of 0.08 to 0.12 [11, 20, 24, 25]. The lower range of C_S value, compared to single phase, could be attributed to the interphase coupling term, which acts as a form of SGS model and can make contribution to the turbulent kinetic energy dissipation. The sensitivity analysis carried out for C_S value shows that larger C_S values can produce excessive damping effect to the liquid velocity field and eventually leads to a steady-state solution [26, 27].

The main reason for the frequent use of the Smagorinsky model is its simplicity. Its drawbacks are that the constant C_S has to be calibrated and its optimal value may vary with the type of flow or the discretization scheme. Moreover, the model is purely dissipative and hence does not account either for the small-scale effect on the large scales adequately (by neglecting the "backscatter" of turbulent energy), while it acts purely as a drain for the turbulent kinetic energy.

The dynamic model, originally proposed by Germano et al. [28], eliminates some of these disadvantages by calculating the Smagorinsky constant as a function of space and time from the smallest scales of the resolved motion.

Dynamic SGS Model

The dynamic SGS model assumes SGS turbulent energy to be in local equilibrium (i.e., production = dissipation). The eddy viscosity is estimated from (7) but with a C_s as a local, time-dependent variable.

The basic idea is to apply a second test filter to the equations. The new filter width, twice the size of the grid filter, produces a resolved flow field. The difference between the two resolved fields is the contribution of the small scales whose size is in between the grid filter and the test filter. The information related to these scales is used to compute the model constant. The advantage here is that no empirical constant is needed and that the procedure allows the negative turbulent viscosity implying energy transfer from smaller to larger scales (energy back-scatter). This effect, in principle, allows both an enhancement and attenuation of the turbulent intensity introduced by the bubbles.

The model has a few drawbacks; wide fluctuations in dynamically computed constants can cause stability issues, along with additional computational expense.

One-Equation Model

In spite of the fact that dynamic SGS model calculates model constant C_s, thus making a constant-free model, it lacks the information on the amount of SGS turbulent kinetic energy, a datum which may prove useful in modelling some aspects of dispersed flows (e.g., SGS bubble-induced turbulence).

The essence of the one-equation model is to solve additional transport equation for SGS turbulent kinetic energy:

$$\frac{\partial k_{SGS}}{\partial t} = \nabla \left[(\mu + \mu_{SGS}) \nabla k_{SGS} \right] + P_{k_{SGS}} - C_\varepsilon \frac{k_{SGS}^{3/2}}{\Delta}.$$

(8)

Here, P_{KSGS} is production of SGS turbulent kinetic energy and is defined as

$$P_{k_{SGS}} = \mu_{SGS} \left| S_{ij} \right|,$$

(9)

and SGS viscosity is obtained from

$$\mu_{SGS} = C_k \Delta k_{SGS}^{1/2}.$$

(10)

The availability of the SGS turbulent kinetic energy allows for modelling of SGS interphase sorces such as bubble-induced turbulence and turbulent dispersion at SGS. The application of one-equation SGS model for bubbly flows is illustrated in more detail in sections below.

Effect of Bubble-Induced Turbulence (BIT)

In the E-E approach, the turbulent stress in the liquid phase is considered to have two contributions, one due to the inherent, that is, shear-induced turbulence that is assumed to be independent of the relative motion of bubbles and liquid and the other due to the additional bubble-induced turbulence (Sato and Sekoguchi [37]). For BIT there are two modelling approaches. The first approach is proposed by Sato and Sekoguchi [37] and Sato et al. [42]:

$$\mu_{BI,l} = \rho_f C_{\mu,BI} \, \alpha_g \, d_b \left| u_g - u_l \right|,$$

(11)

with $C_{\mu,BI}$ as a model constant which is equal to 0.6 and d_b as the bubble diameter. Milelli et al. [11, 24] found that the modelling of the bubble-induced turbulence did not improve the results. They tried two different formulations: the Tran model and the Sato model and found that they have negligible effect. This was attributed to fact that the bubble-induced viscosity (and turbulence) is not crucial, the turbulence being mainly driven by the liquid shear, and a low void fraction ($\approx$2% leading to $\mu_{BI,} \approx 10^{-2}$ kg/(ms)) did not significantly modify the situation. It was thought that in a case in which the bubbles actually drive the turbulence (via buoyancy and/or added mass forces), the situation would be different. However, in subsequent studies, similar observations were made in bubble plumes simulated by Deen et al. [19], Dhotre et al. [20], Ničeno et al. [9].

The second approach for the modelling of BIT allows for the advective and diffusive transport of turbulent kinetic energy. This model incorporates the influence of the gas bubbles in the turbulence by means of additional source terms in the k_{SGS} equation and is taken to be proportional to the product of the drag force and the slip velocity between the two phases. This approach was used in work of Niceno et al. [10] through the use of a one-equation model. They found significant influence of the additional source terms as used by Pfleger et al. [43], as shown in Figure 2.

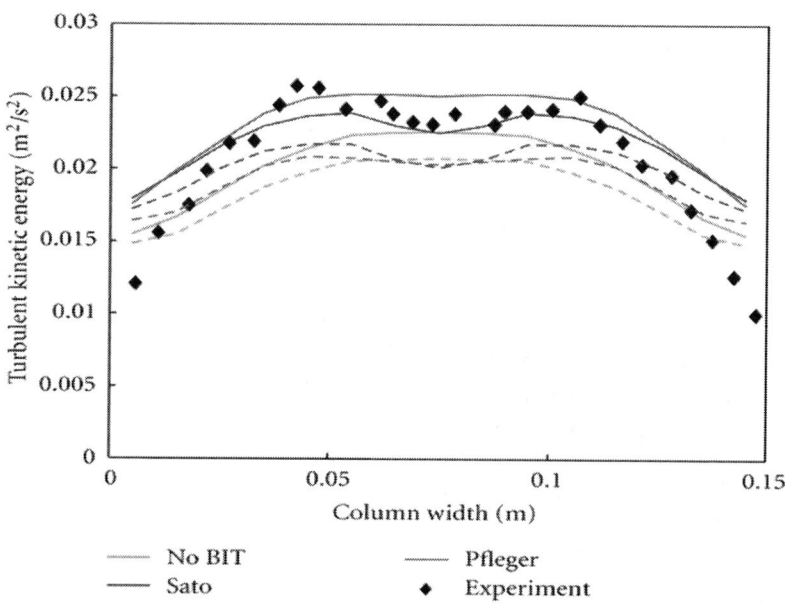

(a)

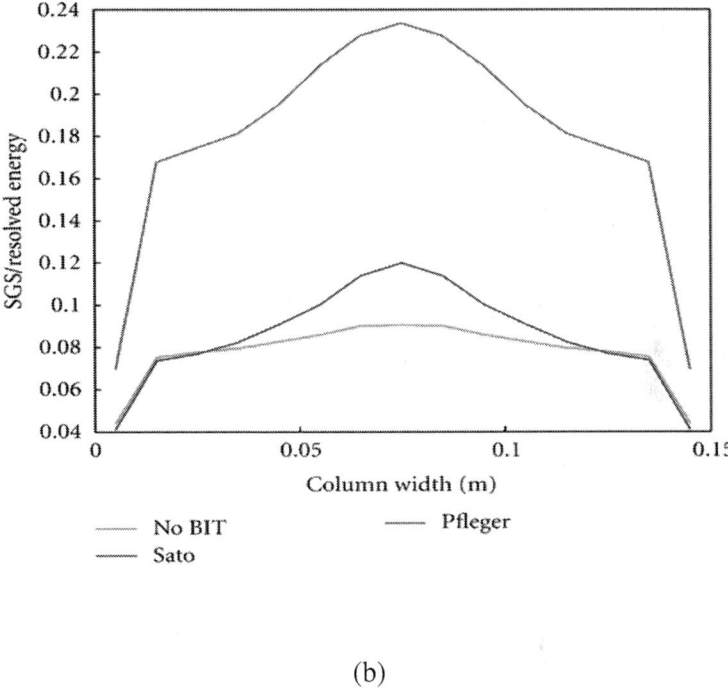

(b)

Figure 2: (a) Resolved (dashed) and total (continuous) liquid kinetic energy and (b) ratio of the modelled and resolved parts of the turbulent kinetic for various BIT models. (from Niceno et al. [9]).

Figure 2 shows the comparison of the liquid kinetic energy obtained for the case of a bubble plume rising in a square column. It can be seen that the simulation without BIT underpredicts the turbulent kinetic energy. The use of the Sato model reproduced the double-peaked profile for kinetic energy. The Pfleger model also reproduced the experimental data very well. Figure 2(b) shows the ratio of the modelled SGS energy to the resolved energy. With no BIT, this ratio has the lowest value, whereas the Sato model yields more SGS energy, while the Pfleger model gives a ratio that is roughly twice as high, which is particularly pronounced in the middle of the column. Table 2 gives a summary of BIT models proposed by various investigators.

Table 2: Bubble-induced turbulence models

No.	Author	μ_{BIT}	$S_{k,BIT}$	$S_{\varepsilon,BIT}$	Assumptions
(1)	Sato and Sekoguchi [37]	$\mu_{BIT} = \rho_L \alpha_G$ $C_{\mu,BIT} d_B |U_G - U_L|$	0	0	
(2)	Pfleger and Becker [38]	0	α_L $C_K |M_K||U_G - U_L|$	$\dfrac{\alpha_L}{K_L} C_\varepsilon S_{k,BIT}$	
(3)	Troshko and Hassan [39]	0	$|M_{D,L}||U_G - U_L|$	$0.45 \dfrac{3C_D |U_G - U_L|}{2C_{VM}d_B} S_{K,BIT}$	
(4)	Crowe et al.[14]				PSI cell/ball approximation
(5)	Sommerfeld [40]				Stochastic interparticle collision model
(6)	Sommerfeld et al. [41]				Langevin equation model

Table 3: Drag force models

No.	Author	Equation
(1)	Ishii and Zuber [44]	$C_D = \dfrac{24}{Re}(1 + 0.1Re^{0.75})$
(2)	Tomiyama [45]	$C_D = \dfrac{(8/3)Eo(1-E^2)}{E^{2/3}Eo + 16(1-E^2)E^{4/3}}F(E)^{-2}$
		$E = \dfrac{1}{1+0.163Eo^{0.757}}$ (Wellek et al. [46])
		$F(E) = \dfrac{\sin^{-1}\sqrt{1-E^2} - E\sqrt{1-E^2}}{(1-E^2)}$
(3)	Tomiyama [47] (pure system)	$C_D = \max[\min[\dfrac{16}{Re}(1+0.15Re^{0.687}),\dfrac{48}{Re}],\dfrac{8}{3}\dfrac{Eo}{Eo+4}]$

(4)	Ishii and Zuber [44] (distorted regime)	$C_D = \dfrac{2}{3} Eo^{1/2}$
		$Eo = g\Delta\rho d_c^2/\sigma$
(5)	Clift et al. [48]	$C_D = \dfrac{24}{Re}(1+0.15 Re_p^{0.687}), R_{ep} \leq 800$
		$0.44\ R_{ep} \leq 800$
(6)	Tomiyama [47] (contaminated system)	$C_D = \max[\min[\dfrac{24}{Re}(1+0.15 Re^{0.687})\dfrac{48}{Re}], \dfrac{8}{3}\dfrac{Eo}{Eo+4}]$

NUMERICAL DETAILS

Crucial parameters for obtaining reliable LES results are the time step selection, the total time for gathering good statistics of the averaged variables, and discretization schemes for the variables. The time step choice is determined by the criterion that the maximum Courant-Fredrichs-Levy (CFL) number must be less than one ($N_{CFL}=\Delta t u_{max}/\Delta x_{max}<1$).

For flow variables, central difference should be used for discretization of advection terms and avoid using diffusive upwind schemes. However, for scalars variables, high-order schemes (MUSCL, QUICK, or Second-Order) may be tolerable to avoid nonphysical solutions (e.g., negative volume fractions). An alternative to high-order schemes are the bounded central differences. The risk with use of all but central scheme is their diffusivity. Their influence on LES may exceed the modelled SGS transport.

It is necessary to follow the initial phase of the simulation, wherein the turbulent strutures develop starting from initial condition and to reach a statistiacally steady state. The duration of this phase depends on the flow characteristics. The simulation must be run for a total time long enough to allow all turbulent instabilities that develop during this phase to be convected across the region of interest. However, the convecting velocities of the turbulent structures and the regions of interest are not always known as a priori. This is why it is recommended to run the simulation a multitude (typically 5 times) of the slowest integral time scales, which often is the flow through time defined as the ratio of the system height over the bulk (superficial) velocity.

LES PREDICTION OF THE FLOW PATTERN FOR DISPERSED BUBBLY FLOWS

Here, we review different LES studies that were performed using the E-E and E-L approaches for simulating flow patterns in gas-liquid bubbly flows. Table 1 gives a summary of key numerical parameters (filter size, number of grids, SGS model, bubble diameter, coefficient for interfacial forces) and experimental details (geometrical dimension, sparger design, range of superficial gas velocity) used by investigators.

Euler-Eulerian (E-E) Approach

Milelli et al. [11, 24, 49]

Milelli et al. reported for the first time two-phase LES with E-E approach. They first investigated statistically 2D flow configuration and then free bubble plume.

They addressed important concerns related to the two-phase LES simulation. For instance, they found that the optimum ratio of the cutoff filter width (i.e., the grid) to the bubble diameter (d_b/Δ) should be around 1.5. That means mesh size should be at least 50% larger than the bubble diameter (Figure 1) so that (a) bubble size determines the largest scale modelled (b) and its interaction with the smallest calculated scale above the cut-off is captured. This is also supported by the scale-similarity principle of Bardina et al. [50].

Milelli [49] investigated LES for a free bubble plume and compared their predictions with the experiment of Anagbo and Brimacombe [51]. Here, they found that the mean quantities were not strongly affected by the different SGS models. Moreover they found little impact of the dispersed phase on the liquid turbulence, from the turbulent energy spectrum taken in the bubbly flow region which revealed a power-law distribution oscillating between −5/3 and −8/3 in the inertial subrange. The results conform to previous studies, which attributed the more dissipative spectrum to the presence of the dispersed phase. Hence, they found no influence of modifying the SGS model to account for bubble-induced dissipation.

Further, they observed in simulation that the lift coefficient value plays a major role in capturing the plume spreading and the used lift coefficient may differ for an LES compared to the one that is justified in an RANS approach. The plausible explanation here is from different handling of two factors responsible for bubble dispersion, that is, interaction between the bubbles and influence of turbulent eddies in the liquid phase.

Deen et al. [19]

Deen et al. [19] reported LES for gas-liquid flow in a square cross-sectional bubble column for the first time. They investigated the performance of RANS and LES approaches, influence of the interphase forces, and bubble-induced turbulence.

They found that RANS approach (k-ε model) overestimated the turbulent viscosity and could only predict low frequency unsteady flow. On other hand, LES as shown in Figure 3 reproduced high frequency experimental data and predicted the strong transient bubble plume movements as in an experiment.

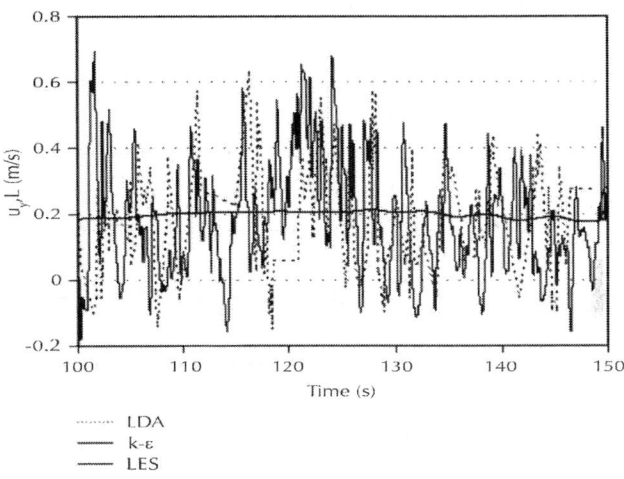

Figure 3: Time history of the axial liquid velocity at the centreline of the column, at a height of 0.25 m (from Deen et al. [19]).

Furthermore, they also identified that the lift force is responsible for transient spreading of the bubble plume and in absence of it, only with drag force, the bubble plume showed no transverse spreading.

They considered the effective viscosity of the liquid phase with three contributions: the molecular, shear-induced turbulent (modelled using Smagorinsky model), and bubble-induced turbulent viscosities [37]. Like in the work of Milelli, they confirmed the marginal effect of the BIT on the predictions. The effect of virtual mass force on the simulated results was also found to be negligible.

Bove et al. [29]

Bove et al. [29] reported LES with E-E approach for the same square cross-sectional bubble column as used by Deen et al. [19]. They studied the influence of numerical modelling of the advection terms and the inlet conditions on LES performance. The upwind first-order and higher-order Flux Corrected Transport (FCT) schemes for both the phase fraction equations and the momentum equations were employed. The simulations using a second-order FCT scheme showed relatively good agreement with the measurement data of Deen et al. [19]. The authors showed that the proper discretization of the momentum and volume fraction equations is essential for correct prediction of the flow field.

Further, the LES results were found to be very sensitive to inlet boundary conditions (Figure 4). Three different inlet configurations simulated showed that the inlet modelling influences the predicted fluid flow velocity (as in Figure 4(a)) and an important fluid flow parameter, the turbulent viscosity (Figure 4(b)). In this work, the sparger (a perforated plate) was not modelled due to the difficulty in adapting the mesh grid to the geometry. They also suggested that near wall region description in the SGS models is important, and the lack of the near wall modelling can lead to erroneous prediction of frictional stresses at the wall.

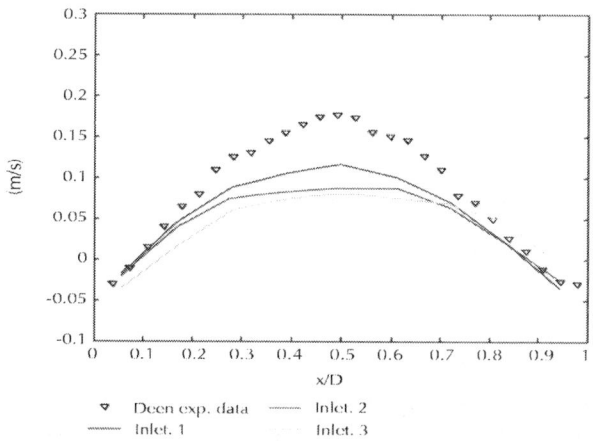

(a)

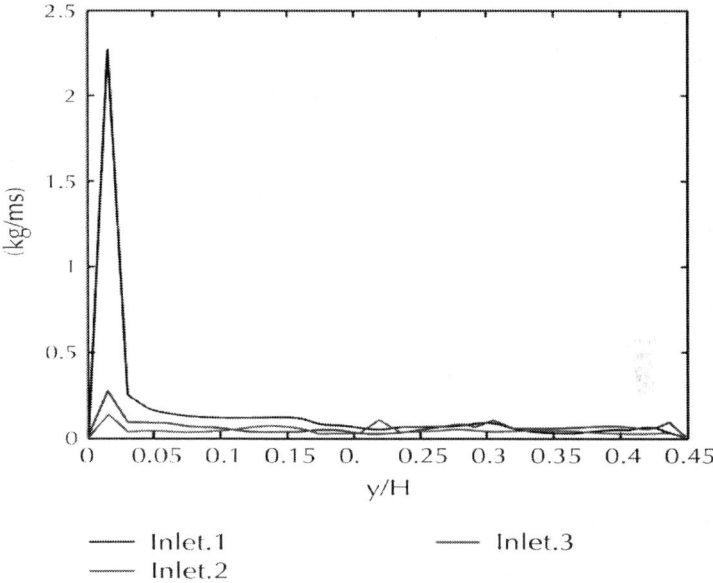

Figure 4: Comparison of (a) averaged axial liquid velocity profile at y/H=0.56, (b) instantaneous viscosity profile along the height of the column (120 s) for three inlet conditions (from Bove et al. [29]).

They used drag model for the contaminated water which gave a better prediction of the slip velocity; however, the velocity profile was underestimated for both gas and liquid phase. Reason for the underprediction was not clear, whether it was due to drag model or an improper value of the lift coefficient used or an error in the near wall modelling. Need for further work in this direction was suggested.

Zhang et al. [26]

Zhang et al. [26] reported LES in a square cross-sectional bubble column. They investigated the Smagorinsky model constant and carried out a sensitivity analysis. It was found that higher C_S values led to higher effective viscosity which dampens the bubble plume dynamics leading to a steep mean velocity profile (as shown in Figure 5). They obtained a good agreement with the measurements with C_S in range of 0.08–0.10. They also confirmed that the lift force plays a critical role for capturing the dynamic behaviour of the bubble plume.

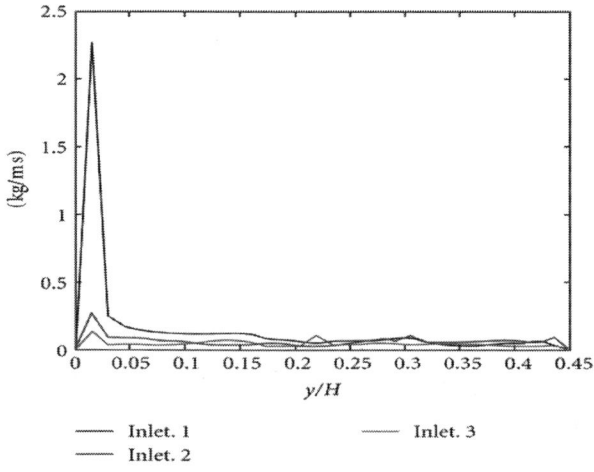

(a)

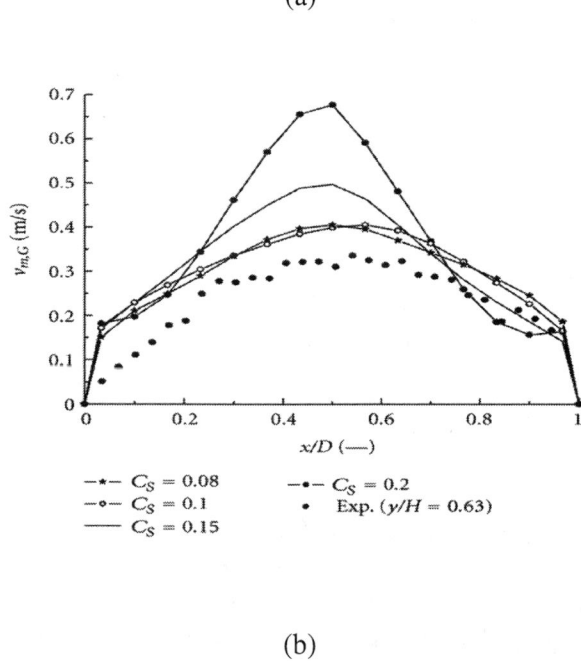

(b)

Figure 5: Comparison of the prediction and measurement of mean velocity of the both phases; the predicted profiles were obtained with different C_s values used in the SGS model (Zhang et al. [26]).

They extended the work of Deen et al. [19] and predicted the dynamic behaviour in the square bubble column using a k-ε turbulence model extended with BIT.

Tabib et al. [17]

Tabib et al. [17] reported LES using E-E approach in a cylindrical column for a wide range of superficial gas velocity. In accordance with the earlier work, they confirmed the importance of a suitable lift coefficient and drag law. Moreover, they studied the influence of different spargers (perforated plate, sintered plate, and single hole) and turbulence models (k-ε, RSM, and LES) using the experimental data of Bhole et al. [52]. The main findings from the study were that the RSM performs better than the k-ε model; the LES was successful in predicting the averaged flow behaviour and was able to simulate the instantaneous vortical-spiral flow regime in the case of a sieve plate column, as well as the bubble plume dynamics in case of single-hole sparger. Finally, they concluded that LES can be effectively used for the study of the flow structures and instantaneous flow profiles.

Dhotre et al. [20]

Dhotre et al. [20] reported LES with an E-E approach for a gas-liquid flow in a square cross-sectional bubble column. They studied the influence of SGS models: Smagorinky and Dynamic models of Germano et al. [28]. It was found that both the Smagorinsky model ($C_s = 0.12$) and the Germano model predictions compared well with the measurements.

They further investigated the value of C_s obtained from the Germano model. Reason for similar performance of both models was clear from the probability density function of C_s (from Germano model) over the entire column. As shown in Figure 6, the value of C_s has the highest probability in the range of 0.12–0.13. Like Zhang et al. [26], the authors confirmed that with a proper BIT model, RANS also performed well for mean quantities of flow variables. Figure 7 shows the comparison of the predicted instantaneous vector flow field for axial liquid velocity from all the three models (Smagorinky, Germano and RANS).

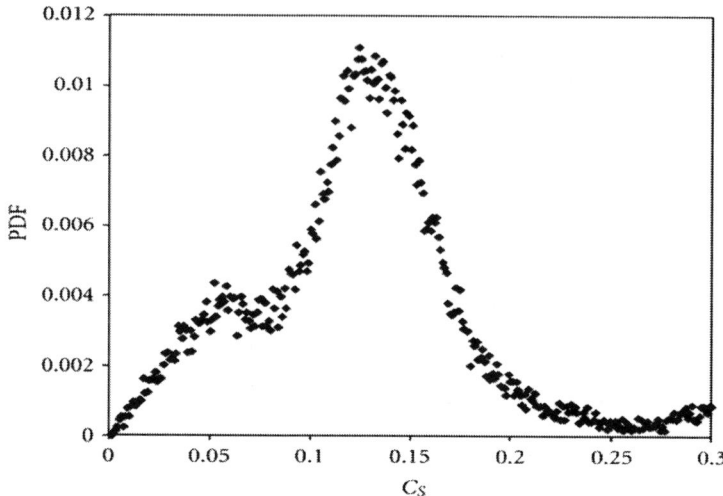

Figure 6: Probability density function for computed constant C_S in Germano model over entire column (from Dhotre et al. [9]).

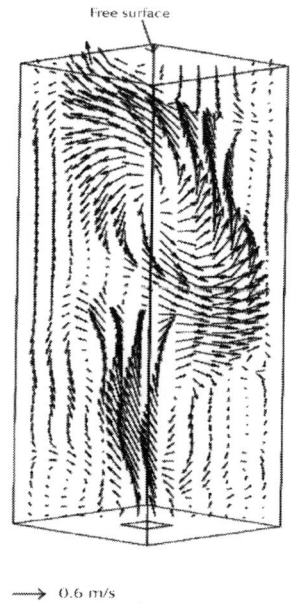

(a)

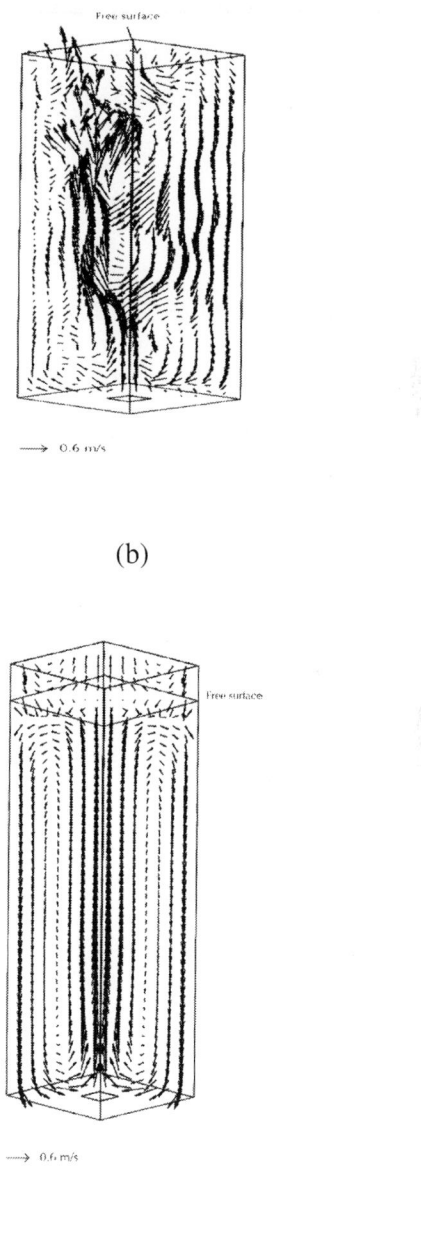

(b)

(c)

Figure 7: Predicted instantaneous vector flow field for axial liquid velocity after 150 s, for all three models (from Dhotre et al. [20]).

It was further concluded that the Germano model can give correct C_S estimates for the configuration under consideration and, in general, can be used for other systems where C_S is not known as "a priori" from previous analysis.

Niceno et al. [10]

Niceno et al. [10] investigated LES with E-E approach for a gas-liquid flow in a square cross-sectional bubble column. They demonstrated the applicability of a one-equation model for the SGS kinetic energy (K_{SGS}). The predictions showed that the one-equation SGS model gives superior results to the Germano model with the additional benefit of having information on the modelled SGS kinetic energy:

$$\mu_{\text{eff},l} = \mu_{\text{lam},l} + \rho_l C_k \Delta \sqrt{k_{\text{SGS}}},$$

(12)

with $C_K = 0.07$ a model constant. They studied the influence of two approaches for bubble-induced turbulence: approach of an algebraic model (Sato et al. 1975) and extra source terms (as used in Pflger et al. 1999) in the transport equation for SGS kinetic energy approach. It was found that the latter approach improved the quantitative prediction of the turbulent kinetic energy (as shown in Figure 2(a)). The modelled SGS kinetic energy for the Pfleger model found to be much higher than for the Sato model (Figure 2(b)), indicating the Pfleger model needs a more appropriate constant for LES.

They suggested that the modelled SGS information can be used to access the SGS interfacial forces, in particular the turbulent dispersion force. In their work, the effect of SGS turbulent dispersion force could not be determined as the bubble size was almost equivalent to the mesh size.

Dhotre et al. [18]

Dhotre et al. [18] extended LES with E-E approach for a gas-liquid flow in a large-scale bubble plume. The predictions at three elevations were compared with the measurement data of Simiano [55] and an RANS prediction. The LES approach was shown superior in capturing the transient behaviour of the plume (Figure 8) and predicts second-order statistics of the liquid phase accurately.

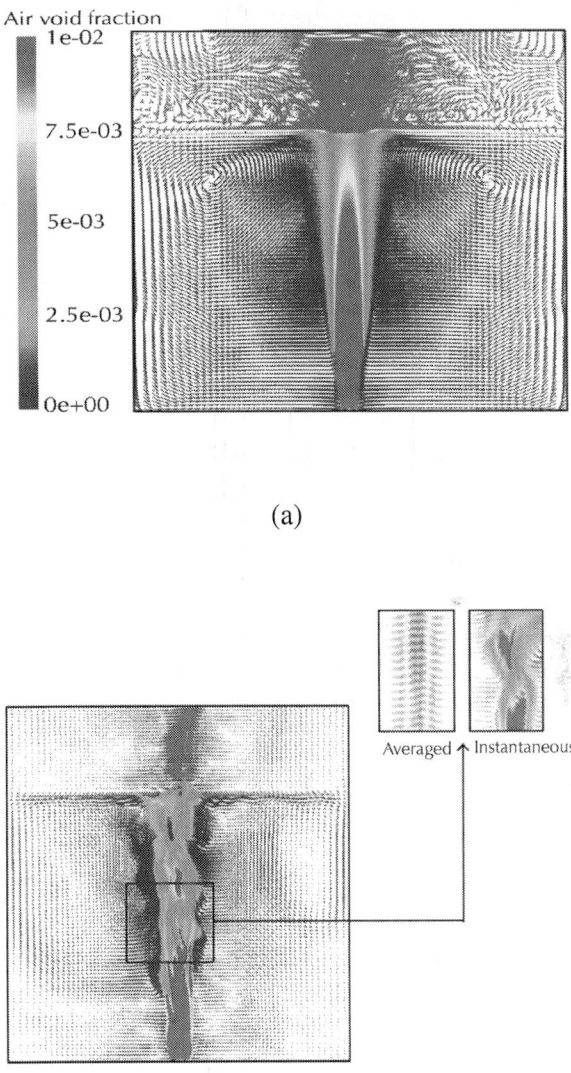

(a)

(b)

Figure 8: Comparison of k-ϵ model and EELES predictions; vector plot of axial velocity coloured with the void fraction in the midplane. (a) k-ϵ model (b) EELES (from Dhotre et al. [18]).

They emphasized the crucial role of the lift force in the prediction of the lateral behaviour of the bubble plumes. In the RANS approach the turbulent dispersion force is required to reproduce the bubble dispersion; however, in LES, bubble dispersion is implicitly calculated by resolving the large-scale turbulent motion responsible for bubble dispersion. The dependence of the bubble dispersion with the value of lift coefficient was also observed in Milelli et al. [11, 24], Deen et al. [19], Lain and Sommerfeld [56], Van den Hengel et al. [31], Tabib et al. (2008), and Dhotre et al. [20]).

Dhotre et al. [18] found good agreement with the measurement data at higher elevation, while discrepancies were observed at lower elevation, near the injector. The reason for the discrepancies was attributed to the absence of modelling bubble coalescence and breakup. This was also found in the work of Van den Hengel et al. [31], wherein the authors showed that most of the coalescence occurs in the lower part of the column and recommended to consider bubble size distribution and coalescence and breakup models for reproducing the bubble behaviour near the sparger.

Niceno et al. [10]

Niceno et al. [10] reported LES with E-E approach for a gas-liquid flow in a square cross-sectional bubble column. They compared two different codes (CFX-4 and Neptune) and two subgrid-scale models (as in Figure9). The prediction from the Smagorinsky model in the Neptune CFD code and the one-equation model of CFX-4 was compared with the measurement data of Deen et al. [19]. Agreement between the predictions from the two SGS models was found to be good, and it was concluded that the influence of the SGS model was small. This is in contradiction with earlier work of Van den Hengel et al. [31], where they showed significant contribution of the SGS model (Figure 10), which is discussed in more detail in section (4.2). It remains to be seen if this was due to the fine mesh used by the authors ($\Delta/d_b=1.2$). Niceno et al. [10] argued that with the known flow pattern in a bubble column, that is, a dominant bubble plume meandering between the confining walls, the biggest eddy having most energy is of the size of the domain cross section. Thus, the grid used in their work was a compromise between sufficiently fine to capture the most energetic eddies, and sufficiently coarse to stay close to the Milelli criterion [11,

24]. Furthermore, they pointed out the limitations of LES with E-L or E-E approach without resolving interface; they indicated that the most influential interfacial forces (drag and lift) are modelled for the large-scale field and their effect from the small scale remains a question. On the other hand, they recommend large-scale simulation, as in the works of Lakehal et al. [25], which explicitly resolves the large-scale part of the interfacial forces and models the part at the SGS level, where the effects are smaller and hence less influential on the accuracy of the results.

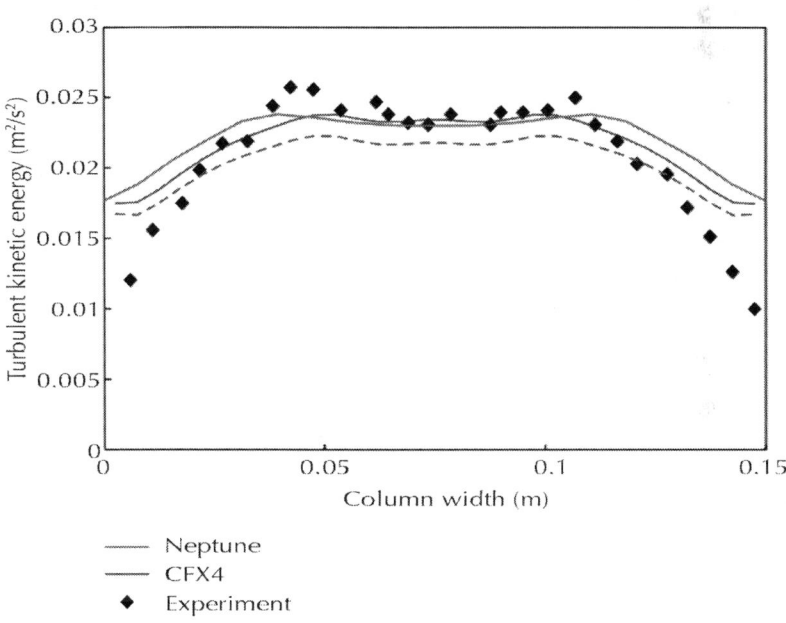

Figure 9: Comparison of liquid turbulent kinetic energy obtained with CFX-4 using one-equation model and Neptune CFD with Smagorinsky model and experimental data. The blue dashed line is the resolved, the blue continuous line is the total (resolved plus SGS) kinetic energy (from Niceno et al. [10]).

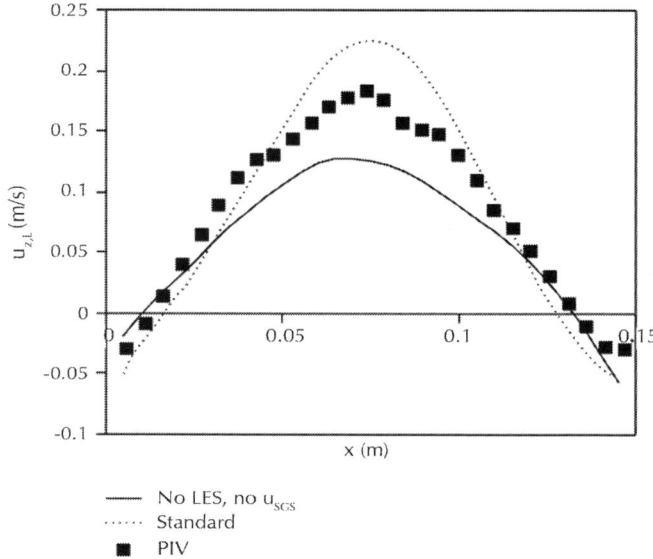

(a)

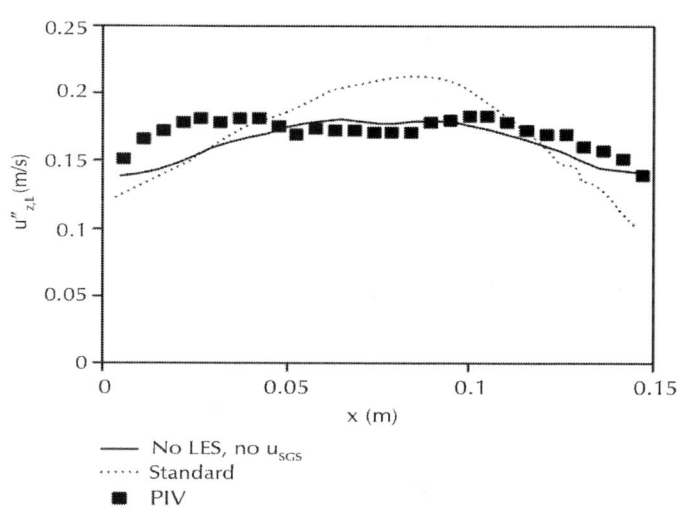

(b)

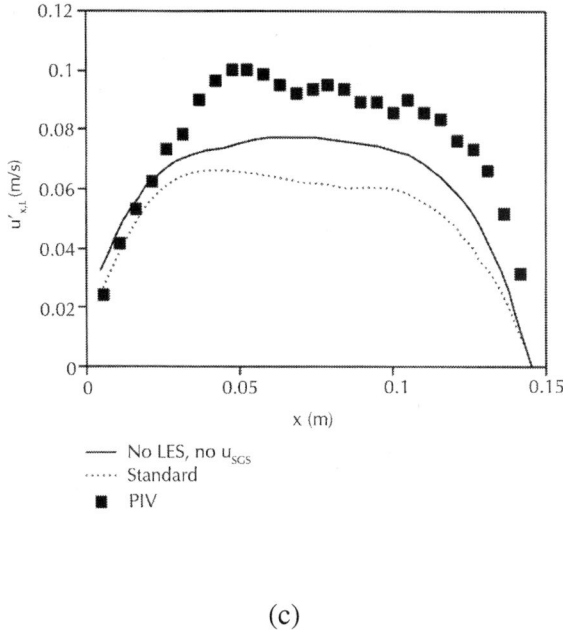

(c)

Figure 10: Comparison of the simulated and experimental liquid velocity and velocity fluctuations for cases with and without SGS model at a height of 0.255 m and a depth of 0.075 m. Effect of the SGS model (from Van den Hengel et al. [31]).

Tabib and Schwarz [30]

Tabib and Schwarz [30] extended the work of Niceno et al. [9] and attempted to quantify the effect of SGS turbulent dispersion force for different particle systems, where the particle sizes would be smaller than the filter size. They used LES with E-E approach.

They used the formulation of Lopez de Bertodano [57] to approximate the turbulent diffusion of the bubbles by the SGS liquid eddies for a gas-liquid bubble column system [17]. The bubble size was in range of 3–5 mm. The mesh used in simulations was coarser than the bubble diameter. They found a high contribution from the SGS turbulent dispersion force, when compared with the magnitude of the other interfacial forces (like drag force, lift force, resolved turbulent dispersion force, and force due to momentum advection and pressure).

Finally, Tabib and Schwarz concluded that for LES with E-E approach, when the mesh size is bigger than bubble size, the SGS turbulent dispersion force should be used, and a one-equation SGS-TKE model overcomes a conceptual drawback of E-E LES model.

Euler-Lagrangian (E-L) Approach

Van den Hengel et al. [31]

Van den Hengel et al. [31] reported LES with E-L approach for a gas-liquid flow in a square cross-sectional bubble column. The liquid phase was computed using LES, and a Lagrangian approach was used for the dispersed phase. They used a discrete bubble model (DBM) originally developed by Delnoij et al. [58, 59] and extended it to incorporate models describing bubble breakup and coalescence. The mean and fluctuating velocities predicted in the simulations showed a good agreement with the experimental data of Deen et al. [19].

Authors studied the influence of the SGS model on the predictions and found that without SGS model, the average liquid velocity and liquid velocity fluctuations are much lower compared to the case with a SGS model. This was due to the lower effective viscosity in this case, which led to less dampening of the bubble plume dynamics and subsequently to flatter mean liquid velocity profiles (as shown in Figure 10).

In this work also, the authors confirmed the important role of the lift coefficient in capturing the plume dynamics. They considered two lift coefficients (C_L= 0.5 and 0.3) and found that a smaller value of the lift coefficient led to higher average velocity and velocity fluctuations and less spreading of the plume, which resulted in overprediction of the average velocity in the centre of the column.

Hu and Celik [32]

Hu and Celik [32] studied LES with an E-L approach for the gas-liquid flow in a flat bubble column. The liquid phase was computed using LES, and a Lagrangian approach was used for the dispersed phase. The authors developed a mapping technique called particle-source-in-ball

(PSI-ball) for coupling the Eulerian and Lagrangian reference frames. The concept is a generalization of the conventional particle-source-in-cell (PSI-cell) method as well as a template-function-based treatment [14].

They reported second-order statistics of the pseudo-turbulent fluctuations and demonstrated that a single-phase LES along with a point-volume treatment of the dispersed phase could serve as a viable closure model.

Hu and Celik reported that the predicted mean quantities (such as mean liquid velocity field) were in good agreement with the experimental data of Sokolichin and Eigenberger [54], as shown in Figure 11, and further gave an accurate prediction of the instantaneous flow features, including liquid velocity fluctuations and unsteady bubble dispersion pattern. Hu and Celik also studied the influence of the Smagorinsky constant and found that the constant for multiphase systems falls in a relatively smaller range than for single-phase flows. Higher values of the C_s showed an excessive damping effect to the liquid field, which led to a steady-state solution. This observation is in accordance with other investigators [26, 31]. Furthermore, authors proposed to use C^s as a modeling parameter rather than a phyiscal constant, as the interphase coupling terms used as well as the high frequency turbulent fluctuations contribute to the turbulent kinetic energy dissipation.

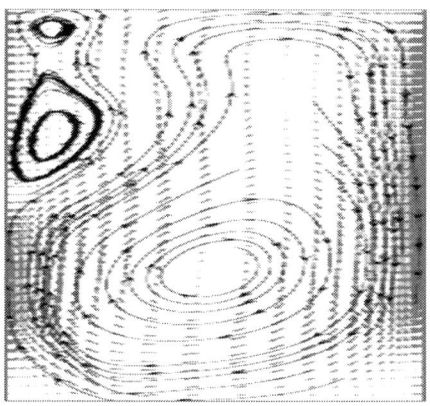

(a)

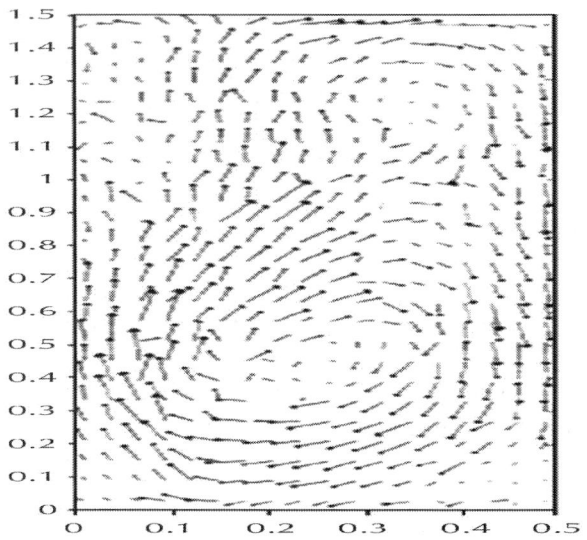

(b)

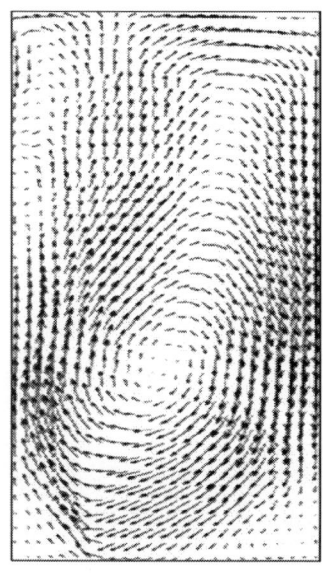

(c)

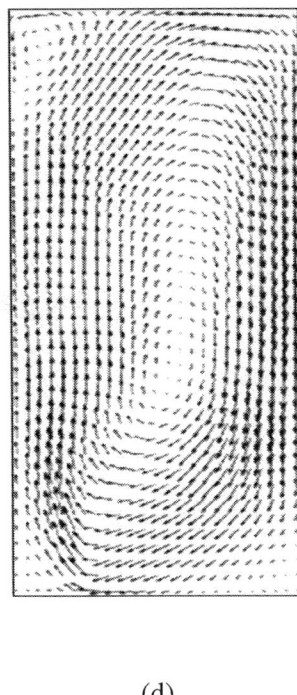

(d)

Figure 11: Long-time averaged liquid velocity field on middepth plane: (a) E-L approach, (b) LDA measurement of Becker et al. [53], (c) LDA measurement of Sokolichin and Eigenberger [54], and (d) 3D E-E simulations of Sokolichin and Eigenberger [54] (from Hu and Celik [32]).

Lain [27]

Lain [27] reported an LES with E-L approach for a gas-liquid flow in a cylindrical bubble column. He used LES for the liquid phase, and a Lagrangian approach for the dispersed gas phase. The interaction terms between liquid and gas phases was calculated using the particle-source-in-cell (PSI-cell) approximation of Crowe et al. [14]. The bubbles were considered as a local source of momentum, and source term was added.

A simple model for the subgrid liquid fluctuating velocity to account for the BIT considered in this work was found to have no influence on the predictions. As in previous works, authors confirmed a strong dependency of the bubble dispersion in the column on the value of transverse lift force coefficient used. He concluded that the lift coefficient depends on the bubble-liquid relative velocity and was the main mechanism responsible for the spreading of bubbles across the column crosssection. He further compared the simulation results with particle image velocimetry (PIV) measurements (Border and Sommerfeld [60]) and k-ε calculations.

Darmana et al. [33]

Darmana et al. [33] used the LES with E-L approach for simulating the gas-liquid flow in a flat bubble column and validated the model with experimental data of Harteveld et al. [61]. They investigated seven sparger designs and their influence on the flow structure. It was found that the model captures the influence of different gas sparging very well (e.g., Figure 12 shows one such case simulated). However, in all cases simulated, authors found systematic overprediction of dispersed phase distribution (25%), which was attributed to an inaccuracy of the drag force and the turbulence model at high gas void fractions.

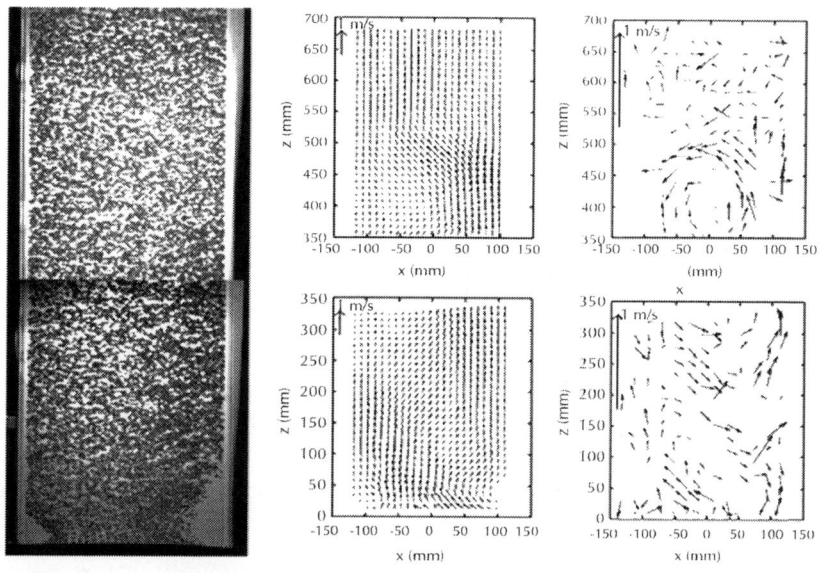

(a)

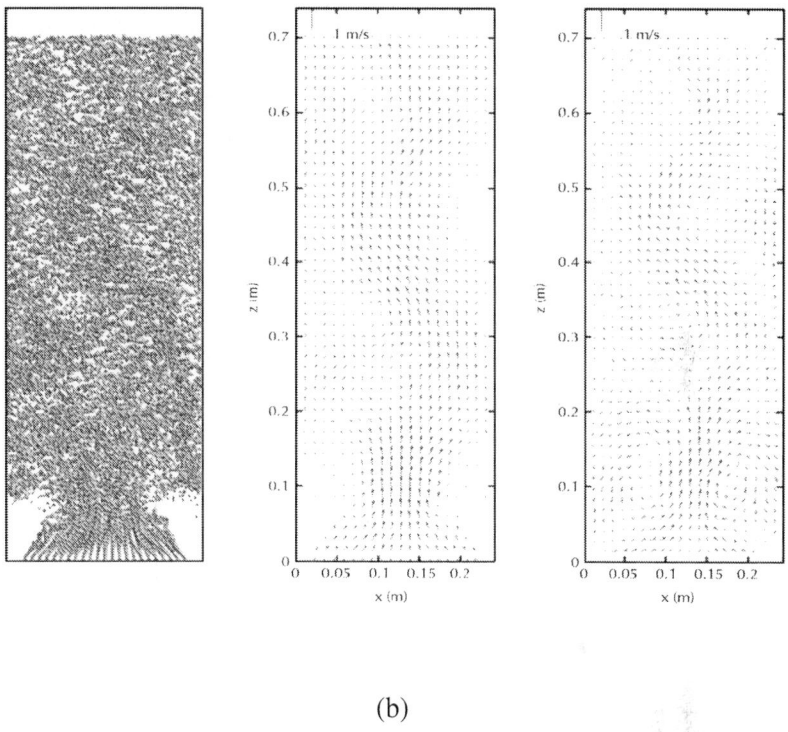

(b)

Figure 12: Instantaneous flow structure comparison between experiment (a) and simulation (b). From left to right: bubble positions, bubble velocity, and liquid velocity (from Darmana et al. [33]).

Sungkorn et al. [34]

Sungkorn et al. [34] reported LES with the E-L approach for a gas-liquid flow in a square cross-sectional bubble column. They modelled the continuous liquid phase using a lattice-Boltzmann (LB) scheme, and a Lagrangian approach was used for the dispersed phase. For the bubble phase, the Langevin equation model [41] was used for estimating the effect of turbulence. The bubble collisions were described by a stochastic interparticle collision model based on the kinetic theory developed by Sommerfeld [40]. The predictions showed a very good agreement with the experimental data for the mean and fluctuating velocity components. Figure 13 shows the sanpshots of predicted the bubble dispersion patterns.

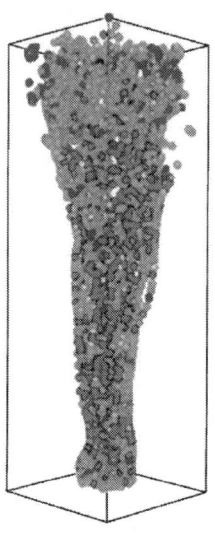

(a)

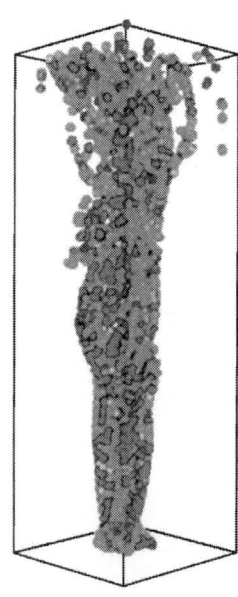

(b)

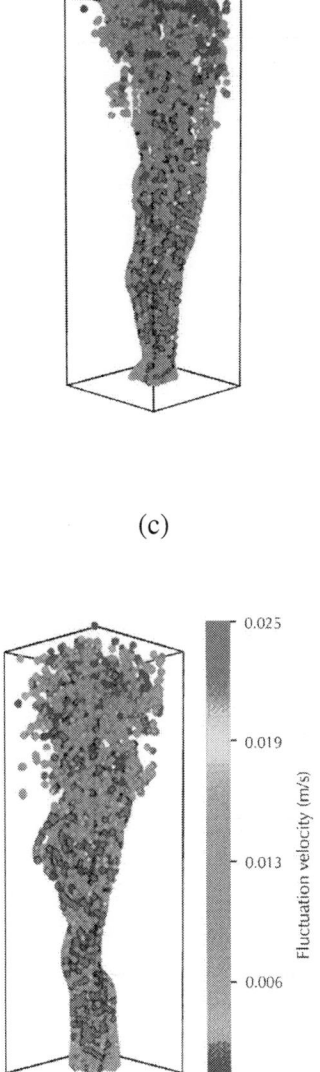

(c)

(d)

Figure 13: Snapshots of the bubble dispersion pattern after 20, 50, 100, and 150 s. The bubbles are coloured by the local magnitude of the liquid fluctuations (from Sungkorn et al. [34]).

It was also found that their collision model leads to two benefits: the computing time is dramatically reduced compared to the direct collision method and secondly it also provides an excellent computational efficiency on parallel platforms. Sungkorn et al. [34] claim that the methodology can be applied to a wide range of problems. The investigations are valid for lower global void fraction, and further work is required to consider it for higher void fraction systems.

APPLICATION OF LES

Preamble

The investigations discussed in earlier sections dealt with the use of LES for predicting the flow patterns. In the published literature, the knowledge of flow pattern has been employed for the estimation of equipment performance such as mixing (Joshi and Sharma [62], Joshi [63], Ranade and Joshi [64], Ranade et al. [65], and Kumaresan and Joshi [66]), heat transfer (Joshi et al. [67], Dhotre and Joshi [68]), Sparger design (Dhotre et al. [69], Kulkarni et al. [70]), gas induction (Joshi and Sharma [71], Murthy et al. [72]), and solid suspension (Raghava Rao et al. [73], Rewatkar et al. [74], and Murthy et al. [75]). Joshi and Ranade [76] have discussed the perspective of computational fluid dynamics (CFD) in designing process equipment with their views on expectations, current status, and path forward. The LES simulations provide substantially improved understanding of the flow pattern. Therefore, in this section, the application of LES for design objectives like mixing, heat transfer, and chemical reactions by some investigators will be reviewed. The LES simulations have also been used in the identification of turbulent structures, their dynamics, and the role of structure dynamics in the estimation of design parameters. The LES simulations have also been used in the estimations of terms ink-ε and RSM models such as generation, dissipation and transport of turbulent kinetic energy (k), the turbulent energy dissipation rate (ε), and Reynolds stresses. These estimations have improved the understanding of RANS (k-ε and RSM) models. These two applications of LES are also described briefly.

Mass Transfer and Chemical Reaction

Darmana et al. [77, 78]

Darmana et al. [77, 78] used LES with E-L approach to simulate flow, mass transfer, and chemical reaction in flat bubble column. They considered mass transfer, rate in liquid-phase momentum equation and reaction interfacial forces in the bubble motion equation.

Also, the presence of various chemical species was accounted through a transport equation for each species. Darmana et al. estimated the mass transfer rate from the information of the individual bubbles directly. They used the model to simulate the reversible two-step reactions found in the chemisorption process of CO_2 in an aqueous NaOH solution in a lab-scale pseudo-2D bubble column reactor (e.g., Figure 14). They found good agreement between simulation and measurement for the case without mass transfer. In absence of an accurate mass transfer closure, the authors found that the overall mass transfer rate was lower compared to the measurement. However, the influence of the mass transfer on the flow agreed well with experimental data.

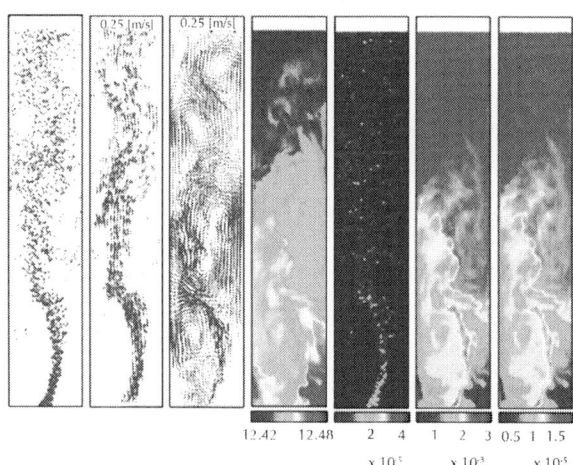

Figure 14: Instantaneous solution 10 s after the CO_2 gas is introduced. From left to right: bubble positions, gas velocity, liquid velocity, pH, respectively,

and concentration of dissolved CO_2, HCO^-, and CO_2^- (kmol m^{-3}) (from Darmana et al. [77]).

Zhang et al. [79]

Zhang et al. [79] followed a procedure similar to that used by Darmana et al. [78], although in this case an E-E approach was used to simulate flow, mass transfer, and chemical reactions in square cross-sectional bubble column [19]. Zhang et al. studied physical and chemical absorption of CO_2 bubbles in water and in an aqueous sodium hydroxide (NaOH) solution. They used a bubble number density equation for coupling of flow, mass transfer, and chemical reaction. The authors demonstrated the influence of the mass transfer and chemical reaction on the hydrodynamics, bubble size distribution, and gas holdup.

Mixing and Dispersion

Bai et al. [36]

Bai et al. [36] used LES with E-L approach to investigate the effect of the gas sparger and gas phase mixing in a square cross-sectional bubble column. The liquid phase was computed using LES, and a Lagrangian approach was used for the dispersed phase. They used the DBM and investigated the effect of two SGS models: Smagorinsky [21] and Vreman [80]. They compared the vertical liquid velocity and turbulent kinetic energy of the liquid phase at three different heights with PIV data and found that the model proposed by Vreman performed better than Smagorinsky model.

They further investigated the effect of the gas sparger properties (sparged area and its location) on the hydrodynamics in a bubble column and characterized the macromixing of the gas phase in the column in terms of an axial dispersion coefficient. They compared the predicted liquid phase dispersion coefficient with the literature correlations as shown in Figure 15. The range of superficial gas velocity investigated in work is low compared to what is common in industrial application. For large-scale reactors at high superficial velocities, Bai et al. recommended to extend the discrete bubble modelling with bubble coalescence and breakup.

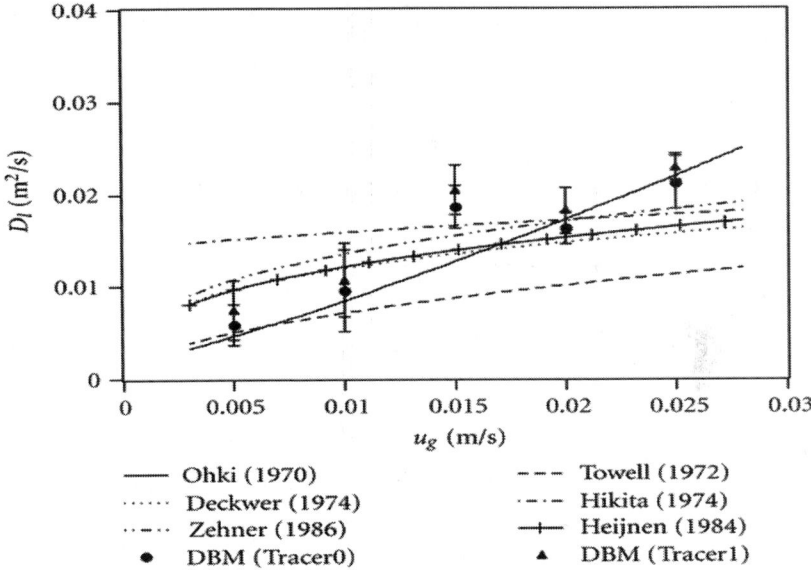

Figure 15: Comparison of the simulated liquid phase dispersion coefficient with the literature correlations (from Bai et al. [36]).

Estimation of the Turbulent Dispersion Force

In the RANS approach, the drag and lift forces depend on the actual relative velocity between the phases, but the ensemble equations of motion for the liquid only provide information regarding the mean flow field. The random influence of the turbulent eddies is considered by modelling a turbulent dispersion force. By analogy with molecular movement, the force is set proportional to the local bubble concentration gradient (or void fraction), with a diffusion coefficient derived from the turbulent kinetic energy. The value of the turbulent dispersion coefficient is chosen to get an agreement with the measurement data and is not known as a priori.

In LES, the resolved part of the turbulent dispersion is implicitly computed, and hence one can use information from LES for calculating the magnitude of this force. The methodology depends on scales at which LES is to be applied. For instance, at the mesoscale, in the E-L approach, bubbles dispersed by drag and lift through turbulent eddies can be computed. At micro-scale LES, one might need to consider bubble coalescence and breakup phenomena along with a reasonable number of bubbles. It can be computationally expensive, but in view of increasing available computer power, this should become feasible soon.

Dynamics of Turbulent Structures and the Estimation of Design Parameters

The turbulent flows contain flow structures with a wide range of length and time scales which control the transport processes. The length scales of these structures can range from column dimensions (highest) to Kolmogorov scales (lowest). However, not all the scales of turbulence contribute equally to different transport rates and mixing. If only mixing is the important design criterion, then the knowledge about the mean flow pattern (large-scale structures) would generally suffice the purpose (Ekambara and Joshi [81]). However, for the prediction of the gas holdup, bubble size distribution, true mass transfer coefficient, and heat transfer coefficient, the knowledge about all the scales is important [82, 83]. Hence, it is imperative to identify the scales and dynamics of turbulent flow structures and their relationship with the rates of different transport process. The present empirical design practices do not consider these basic mechanisms and conceales the detailed local information about the relationship between the turbulence and the equipment performance.

The subject of quantification of local turbulent flow structures and reliable estimation of transport properties has been reviewed by Joshi et al. [84] and [82, 83]. The velocity and pressure data from LES were analyzed using the mathematical techniques such as multiresolution analysis [85], wavelet transforms (discrete and continuous), proper orthogonal decomposition (POD), and hybrid POD-wavelet techniques (Tabib and Joshi [86], Tabib et al. [87], Sathe et al. [88], and Mathpati et al. [89]). These techniques give the size, shape penetration depth, and energy content of all the flow structure in the system. This flow structure information can also be used for the construction of energy spectrum and for examining the scaling laws for turbulence in bubble columns. Such understanding of turbulence is expected to provide better insights into the transport phenomena. One such attempt has been reported by Deshpande et al. [90, 91].

Comparison of Turbulence Models

CFD provides detailed flow information within single- and multiphase reactors. Most popular and computationally inexpensive models such as k-ε model and Reynolds stress model (RSM) are widely used to predict the mean flow pattern. These models can give reliable estimation about the liquid phase mixing. However, they do not accurately predict the turbulence parameters such as turbulent kinetic energy and the dissipation rate due to inbuilt modelling assumptions as well as complexity of flow [11, 24]. These models are time averaged, and hence the information related to different turbulent structures is lost.

It is known that a large number of simplifying assumptions are made while deriving the k-ε and RSM models. Therefore, it is important to understand the gravity of these assumptions on the quantitative values of transport rates of k and ε due to convection, diffusion, and turbulent dispersion. It is also important to know the quantitative estimation of production and dissipation rates of k and ε. Therefore, it is important to estimate these five terms using k-ε, RSM, and LES models. From the LES simulations, the time series of velocity and pressure can be stored. These are subsequently used for the detailed comparison of k-ε, RSM, and LES models in terms of the rates of transport (convection, molecular and turbulent diffusion) and the rates of production, and the dissipation of k and ε for the case of dispersed bubbly flows [92].

SUMMARY AND SUGGESTIONS FOR FUTURE WORK

- E-E and E-L LES are promising approaches for predicting unsteady, buoyancy-driven flow inducing large-scale coherent structures for gas-liquid dispersed flow. Care should be taken to clearly identify the scales (micro, macro, or meso) at which LES should be applied, in order to decide the level of interface resolution and modelling required. The approach of LES at mesoscales (i.e., without explicitly tracking interface) using E-E and E-L description has been reviewed for gas-liquid dispersed flow.

- Pioneering work of Milelli et al. [11, 24] has initiated the LES approach for gas-liquid dispersed flows. The main contribution comes from insights in the cutoff filter requirement and SGS modelling.

- The simulation and the experimental measurement of Deen et al. [19] in a square cross-sectional bubble column have triggered a systematic development of the two-phase LES for both E-E and E-L approaches.

- The concept behind the LES is very simple but characterized by a large number of choices (regarding numerical and physical modelling) that all have significant influence on the results. However, it offers great potential in terms of determination of statistical quantities and instantaneous information about flow structures. This information can be extremely useful for the prediction of other physical processes behaviour (e.g., transport of scalar (temperature, concentration), chemical reactions).

- From LES simulation with E-E/E-L approaches that were reviewed in this work, it is recommended that:

  The grid or filter size selection based on filter size to bubble diameter ratio Δ/d_b of 1.2 gives reasonable results.

  The Smagorinsky constant, C_s, is a modelling parameter rather than a physical constant. Although the constant value of the parameter gives satisfactory results, for unknown configuration, it should be estimated with Germano dynamic procedure (using the overall distribution of the constant through probability density).

  The lift force is the main mechanism for the dispersion, and the lift coefficient should be estimated though sensitivity of interfacial forces on values of slip velocity and gas holdup. The lift coefficient in LES can be different from that in RANS.

  The central difference scheme should be used for the discretization of advection terms for flow variables and high-order schemes (MUSCL, QUICK, or Second-Order) can be used for scalar variables.

  The minimum time for gathering statistics should be at least one flow through time (as defined as ratio of the system height over the bulk (superficial) velocity).

- In advent of computer hardware, the E-L approach appears very promising for the near future. Further work in mapping functions for two-way coupling can expedite the development of this approach that can be used as a means of both predicting the properties of specific turbulent flows and providing flow details that can be used like data to test and refine other turbulence-closure models.

- The approach for BIT with extra production terms into the SGS-turbulent kinetic energy equation (following the procedure described by Pfleger and Becker [38]) has shown to be more effective than the approach involving a bubble-induced viscosity [37]. It can be that the enhanced eddy viscosity in LES does not represented as realistic physical model, as the SGS turbulent kinetic energy. Nonetheless, it is an interesting issue, and more work in investigating the BIT should be undertaken.

- Treatment of the interphase forces needs more attention.

The drag and nondrag forces (lift, virtual mass force) can be modelled using resolved field approaches. The modelling of these forces for the SGS and their effect on the overall simulation results need to be evaluated.

One finds strong dependency of the bubble dispersion on the value of transverse lift force coefficient. The transverse lift, which depends on the bubble-liquid relative velocity, seems to be the main mechanism responsible for the spreading of the bubbles. It will help if one can estimate the separate contributions of each of these forces.

The virtual mass force has little influence on simulation results. So far, a constant coefficient has been used in all the investigations; however, dependence on void fraction has been shown in experiments. It would be good to have a correct description in order to improve results near the inlet where bubble acceleration effects are important.

- The strong coupling between subgrid-scale (SGS) modelling and the truncation error of the numerical discretization can be exploited by developing discretization methods where the truncation error itself functions as an implicit SGS model. Such attempt can be useful and go in the direction of finding a universal SGS model.

- In order to use LES for reliable predictions at minimum computational costs, understanding of the influence of discretization methods, boundary conditions, wall models, and numerical parameters (e.g., convergence criterion, time steps, etc.) is essential. The contribution focusing on these aspects should be undertaken for both E-E/E-L approaches.

- Substantial development has been achieved in LES in the last decade for understanding bubbly gas-liquid dispersed flow. However, it is mainly restricted to low superficial gas velocities and gas fractions. Future work should focus on industrially relevant large-scale reactors at high superficial gas velocity. The modelling of bubble coalescence and breakup might be necessary, along with further clarity in filtering operations.

- Joshi and coworkers have used LES for the identification of flow structures and their dynamics. They have proposed a procedure to use this information for the estimation of design parameters. Substantial additional work is needed for finding 3D information on the structure characteristics such as size, shape, velocity, and energy distributions

ACKNOWLEDGMENTS

N. G. Deen would like to thank the European Research Council for its financial support, under its Starting Investigator Grant scheme, contract number 259521 (cutting bubbles).

REFERENCES

1.   A. Sokolichin, G. Eigenberger, A. Lapin, and A. Lübbert, "Dynamic numerical simulation of gas-liquid two-phase flows: Euler/Euler versus Euler/Lagrange," Chemical Engineering Science, vol. 52, no. 4, pp. 611–626, 1997.

2.   G. Bois, D. Jamet, and O. Lebaigue, "Towards large eddy simulation of two-phase flow with phase-change: direct numerical simulation of a pseudo-turbulent two-phase condensing flow," in Proceedings of the 7th International Conference on Multiphase Flow (ICMF '10), Tampa, Fla, USA, May-June 2010.

3.   A. Toutant, M. Chandesris, D. Jamet, and O. Lebaigue, "Jump conditions for filtered quantities at an under-resolved

discontinuous interface—part 1: theoretical development," International Journal of Multiphase Flow, vol. 35, no. 12, pp. 1100–1118, 2009.

4. A. Toutant, M. Chandesris, D. Jamet, and O. Lebaigue, "Jump conditions for filtered quantities at an under-resolved discontinuous interface—part 2: a priori tests," International Journal of Multiphase Flow, vol. 35, no. 12, pp. 1119–1129, 2009.

5. S. Magdeleine, B. Mathieu, O. Lebaigue, and C. Morel, "DNS upscaling applied to volumetric interfacial area transport equation," in Proceedings of the 7th International Conference on Multiphase Flow (ICMF '10), p. 12, Tampa, Fla, USA, May-June 2010.

6. D. Lakehal, "LEIS for the prediction of turbulent multi-fluid flows applied to thermal hydraulics applications," in Proceedings of the XFD4NRS, Grenoble, France, September 2008.

7. D. Lakehal, M. Fulgosi, S. Banerjee, and G. Yadigaroglu, "Turbulence and heat exchange in condensing vapor-liquid flow," Physics of Fluids, vol. 20, no. 6, Article ID 065101, 2008.

8. D. Bestion, "Applicability of two-phase CFD to nuclear reactor thermalhydraulics and elaboration of best practice guidelines," Nuclear Engineering and Design, vol. 253, pp. 311–321, 2012.

9. B. Ničeno, M. T. Dhotre, and N. G. Deen, "One-equation subgrid scale (SGS) modelling for Euler-Euler large eddy simulation (EELES) of dispersed bubbly flow," Chemical Engineering Science, vol. 63, no. 15, pp. 3923–3931, 2008.

10. B. Niceno, M. Boucker, and B. L. Smith, "Euler-Euler large eddy simulation of a square cross-sectional bubble column using the Neptune CFD code," Science and Technology of Nuclear Installations, vol. 2009, Article ID 410272, 2009.

11. M. Milelli, B. L. Smith, and D. Lakehal, "Large-eddy simulation of turbulent shear flows laden with bubbles," in Direct and Large-Eddy Simulation IV, B. J. Geurts, R. Friedrich, and O. Metais, Eds., pp. 461–470, Kluwer Academic Publishers, Amsterdam, The Netherlands, 2001.

12. D. A. Drew, "Averaged field equations for two-phase media," Studies in Applied Mathematics, vol. 50, no. 2, pp. 133–165, 1971.

13. S. Elgobashi, "Particle-laden turbulent flows: direct simulation and closure models," Applied Scientific Research, vol. 48, no. 3-4, pp. 301–314, 1991.

14. C. T. Crowe, M. P. Sharma, and D. E. Stock, "The particle-source-in cell (PSI-CELL) model for gas-droplet flows," Journal of Fluids Engineering, vol. 99, no. 2, pp. 325–332, 1977.

15. G. Hu, Towards large eddy simulation of dispersed gas-liquid two-phase turbulent flows [Ph.D. thesis], Mechanical and Aerospace Engineering Department, West Virginia University, Morgantown, WVa, USA, 2005.

16. N. G. Deen, M. V. S. Annaland, and J. A. M. Kuipers, "Multi-scale modeling of dispersed gas-liquid two-phase flow," Chemical Engineering Science, vol. 59, pp. 1853–1861, 2004.

17. M. V. Tabib, S. A. Roy, and J. B. Joshi, "CFD simulation of bubble column—an analysis of interphase forces and turbulence models," Chemical Engineering Journal, vol. 139, no. 3, pp. 589–614, 2008.

18. M. T. Dhotre, B. Niceno, B. L. Smith, and M. Simiano, "Large-eddy simulation (LES) of the large scale bubble plume," Chemical Engineering Science, vol. 64, no. 11, pp. 2692–2704, 2009.

19. N. G. Deen, T. Solberg, and B. H. Hjertager, "Large eddy simulation of the gas-liquid flow in a square cross-sectioned bubble column," Chemical Engineering Science, vol. 56, no. 21-22, pp. 6341–6349, 2001.

20. M. T. Dhotre, B. Niceno, and B. L. Smith, "Large eddy simulation of a bubble column using dynamic sub-grid scale model," Chemical Engineering Journal, vol. 136, no. 2-3, pp. 337–348, 2008.

21. J. Smagorinsky, "General circulation experiments with the primitive equations," Monthly Weather Review, vol. 91, pp. 99–165, 1963.

22. P. Moin and J. Kim, "Numerical investigations of turbulent channel flow," Journal of Fluid Mechanics, vol. 118, pp. 341–377, 1982.

23. W. Jones and M. Wille, "Large eddy simulation of a jet in a cross flow," in Proceedings of the 10th Symposium on Turbulent Shear Flows, pp. 41–46, The Pennsylvania State University, 1995.

24. M. Milelli, B. L. Smith, and D. Lakehal, "Large-eddy simulation of turbulent shear flows laden with bubbles," in Direct and Large-

Eddy Simulation IV, B. J. Geurts, R. Friedrich, and O. Metais, Eds., pp. 461–470, Kluwer Academic Publishers, Amsterdam, The Netherlands, 2001.

25. D. Lakehal, B. L. Smith, and M. Milelli, "Large-eddy simulation of bubbly turbulent shear flows," Journal of Turbulence, vol. 3, pp. 1–20, 2002.

26. D. Zhang, N. G. Deen, and J. A. M. Kuipers, "Numerical simulation of the dynamic flow behavior in a bubble column: a study of closures for turbulence and interface forces," Chemical Engineering Science, vol. 61, no. 23, pp. 7593–7608, 2006.

27. D. Lain, "Dynamic three-dimensional simulation of gas liquid flow in a cylindrical bubble column Latin American," Applied Research, vol. 39, pp. 317–329, 2009.

28. M. Germano, U. Piomelli, P. Moin, and W. H. Cabot, "A dynamic subgrid-scale eddy viscosity model,"Physics of Fluids A, vol. 3, no. 7, pp. 1760–1765, 1991.

29. S. Bove, T. Solbergt, and B. H. Hjertager, "Numerical aspects of bubble column simulations,"International Journal of Chemical Reactor Engineering, vol. 2, no. A1, pp. 1–22, 2004.

30. M. V. Tabib and P. Schwarz, "Quantifying sub-grid scale (SGS) turbulent dispersion force and its effect using one-equation SGS large eddy simulation (LES) model in a gas-liquid and a liquid-liquid system,"Chemical Engineering Science, vol. 66, no. 14, pp. 3071–3086, 2011.

31. E. I. V. van den Hengel, N. G. Deen, and J. A. M. Kuipers, "Application of coalescence and breakup models in a discrete bubble model for bubble columns," Industrial and Engineering Chemistry Research, vol. 44, no. 14, pp. 5233–5245, 2005.

32. G. Hu and I. Celik, "Eulerian-Lagrangian based large-eddy simulation of a partially aerated flat bubble column," Chemical Engineering Science, vol. 63, no. 1, pp. 253–271, 2008.

33. D. Darmana, N. G. Deen, J. A. M. Kuipers, W. K. Harteveld, and R. F. Mudde, "Numerical study of homogeneous bubbly flow: influence of the inlet conditions to the hydrodynamic behavior,"International Journal of Multiphase Flow, vol. 35, no. 12, pp. 1077–1099, 2009.

34. R. Sungkorn, J. J. Derksen, and J. G. Khinast, "Modeling of turbulent gas-liquid bubbly flows using stochastic Lagrangian model and lattice-Boltzmann scheme," Chemical Engineering Science, vol. 66, no. 12, pp. 2745–2757, 2011.

35. W. Bai, N. G. Deen, and J. A. M. Kuipers, "Numerical analysis of the effect of gas sparging on bubble column hydrodynamics," Industrial and Engineering Chemistry Research, vol. 50, no. 8, pp. 4320–4328, 2011.

36. W. Bai, N. G. Deen, and J. A. M. Kuipers, "Numerical investigation of gas holdup and phase mixing in bubble column reactors," Industrial & Engineering Chemistry Research, vol. 51, no. 4, pp. 1949–1961, 2012.

37. Y. Sato and K. Sekoguchi, "Liquid velocity distribution in two-phase bubble flow," International Journal of Multiphase Flow, vol. 2, no. 1, pp. 79–95, 1975.

38. D. Pfleger and S. Becker, "Modelling and simulation of the dynamic flow behaviour in a bubble column," Chemical Engineering Science, vol. 56, no. 4, pp. 1737–1747, 2001.

39. A. A. Troshko and Y. A. Hassan, "A two-equation turbulence model of turbulent bubbly flows," International Journal of Multiphase Flow, vol. 27, no. 11, pp. 1965–2000, 2001.

40. M. Sommerfeld, "Validation of a stochastic Lagrangian modelling approach for inter-particle collisions in homogeneous isotropic turbulence," International Journal of Multiphase Flow, vol. 27, no. 10, pp. 1829–1858, 2001.

41. M. Sommerfeld, G. Kohnen, and M. Rueger, "Some open questions and inconsistencies of Lagrangian particle dispersion models," in Proceedings of the 9th Symposium on Turbulent Shear Flows, Kyoto, Japan, 1993, paper no. 15-1.

42. Y. Sato, M. Sadatomi, and K. Sekoguchi, "Momentum and heat transfer in two-phase bubble flow-I. Theory," International Journal of Multiphase Flow, vol. 7, no. 2, pp. 167–177, 1981.

43. D. Pfleger, S. Gomes, N. Gilbert, and H. G. Wagner, "Hydrodynamic simulations of laboratory scale bubble columns fundamental studies of the Eulerian-Eulerian modeling approach," Chemical Engineering Science, vol. 54, no. 21, pp. 5091–5099, 1999.

44. M. Ishii and N. Zuber, "Drag coefficient and relative velocity in bubbly, droplet or particulate flows,"AIChE Journal, vol. 25, no. 5, pp. 843–855, 1979.

45. A. Tomiyama, "Drag lift and virtual mass forces acting on a single bubble," in Proceedings of the 3rd International Symposium on Two-Phase Flow Modeling and Experimentation, Pisa, Italy, September 2004.

46. R. M. Wellek, A. K. Agrawal, and A. H. P. Skelland, "Shape of liquid drops moving in liquid media,"AIChE Journal, vol. 12, no. 5, pp. 854–862, 1966.

47. A. Tomiyama, "Struggle with computional bubble dynamics," in Proceedings of the 3rd International Conference on Multi-Phase Flow (ICMF ‹98), Lyon, France, June 1998.

48. R. Clift, J. R. Grace, and M. E. Weber, Bubbles, Drops and Particles, Academic Press, New York, NY, USA, 1978.

49. M. Milelli, A numerical analysis of confined turbulent bubble plume [Diss. EH. no. 14799], Swiss Federal Institute of Technology, Zurich, Switzerland, 2002.

50. J. Bardina, J. H. Ferziger, and W. C. Reynolds, "Improved subgrid models for large eddy simulation," AIAA paper 80-1358, 1980.

51. P. E. Anagbo and J. K. Brimacombe, "Plume characteristics and liquid circulation in gas injection through a porous plug," Metallurgical Transactions B, vol. 21, no. 4, pp. 637–648, 1990.

52. M. R. Bhole, S. Roy, and J. B. Joshi, "Laser doppler anemometer measurements in bubble column: effect of sparger," Industrial and Engineering Chemistry Research, vol. 45, no. 26, pp. 9201–9207, 2006.

53. S. Becker, A. Sokolichin, and G. Eigenberger, "Gas-liquid flow in bubble columns and loop reactors—part II: comparison of detailed experiments and flow simulations," Chemical Engineering Science, vol. 49, no. 24, part 2, pp. 5747–5762, 1994.

54. A. Sokolichin and G. Eigenberger, "Applicability of the standard k-ε turbulence model to the dynamic simulation of bubble columns—part I: detailed numerical simulations," Chemical Engineering Science, vol. 54, no. 13-14, pp. 2273–2284, 1999.

55. M. Simiano, Experimental investigation of large-scale three dimensional bubble plume dynamics [Dissertation no. 16220], Swiss Federal Institute of Technology, Zurich, Switzerland, 2005.

56. S. Lain and M. Sommerfeld, "LES of gas-liquid flow in a cylindrical laboratory bubble column," inProceedings of the 5th International Conference on Multiphase Flow (ICMF ‹04), Yokohama, Japan, 2004, paper no. 337.

57. M. Lopez de Bertodano, Turbulent bubbly two-phase flow in a triangular duct [Ph.D. thesis], Rensselaer Polytechnic Institute, Troy, NY, USA, 1992.

58. E. Delnoij, F. A. Lammers, J. A. M. Kuipers, and W. P. M. van Swaaij, "Dynamic simulation of dispersed gas-liquid two-phase flow using a discrete bubble model," Chemical Engineering Science, vol. 52, no. 9, pp. 1429–1458, 1997.

59. E. Delnoij, J. A. M. Kuipers, and W. P. M. Van Swaaij, "A three-dimensional CFD model for gas-liquid bubble columns," Chemical Engineering Science, vol. 54, no. 13-14, pp. 2217–2226, 1999.

60. D. Bröder and M. Sommerfeld, "An advanced LIF-PLV system for analysing the hydrodynamics in a laboratory bubble column at higher void fractions," Experiments in Fluids, vol. 33, no. 6, pp. 826–837, 2002.

61. W. K. Harteveld, J. E. Julia, R. F. Mudde, and H. E. A. van den Akker, "Large scale vortical structures in bubble columns for gas fractions in the range of 5–25%," in Proceedings of the 16th International Congress of Chemical and Process Engineering (CHISA ‹04), Prague, Czech Republic, 2004.

62. J. B. Joshi and M. M. Sharma, "A circulation cell model for bubble columns," Transactions of the Institution of Chemical Engineers, vol. 57, no. 4, pp. 244–251, 1979.

63. J. B. Joshi, "Axial mixing in multiphase contactors—a unified correlation," Transactions of the Institution of Chemical Engineers, vol. 58, no. 3, pp. 155–165, 1980.

64. V. V. Ranade and J. B. Joshi, "Flow generated by pitched blade turbines. 1. Measurements using laser Doppler anemometer," Chemical Engineering Communications, vol. 81, pp. 197–224, 1989.

65. V. V. Ranade, J. R. Bourne, and J. B. Joshi, "Fluid mechanics and blending in agitated tanks," Chemical Engineering Science, vol. 46, no. 8, pp. 1883–1893, 1991.

66. T. Kumaresan and J. B. Joshi, "Effect of impeller design on the flow pattern and mixing in stirred tanks,"Chemical Engineering Journal, vol. 115, no. 3, pp. 173–193, 2006.

67. J. B. Joshi, M. M. Sharma, Y. T. Shah, C. P. P. Singh, M. Ally, and G. E. Klinzing, "Heat transfer in multiphase contactors," Chemical Engineering Communications, vol. 6, no. 4-5, pp. 257–271, 1980.

68. M. T. Dhotre and J. B. Joshi, "Two-dimensional CFD model for the prediction of flow pattern, pressure drop and heat transfer coefficient in bubble column reactors," Chemical Engineering Research and Design, vol. 82, no. 6, pp. 689–707, 2004.

69. M. T. Dhotre, K. Ekambara, and J. B. Joshi, "CFD simulation of sparger design and height to diameter ratio on gas hold-up profiles in bubble column reactors," Experimental Thermal and Fluid Science, vol. 28, no. 5, pp. 407–421, 2004.

70. A. V. Kulkarni, S. V. Badgandi, and J. B. Joshi, "Design of ring and spider type spargers for bubble column reactor: experimental measurements and CFD simulation of flow and weeping," Chemical Engineering Research and Design, vol. 87, no. 12, pp. 1612–1630, 2009.

71. J. B. Joshi and M. M. Sharma, "Mass transfer and hydrodynamic characteristics of gas inducing type of agitated contactors," Canadian Journal of Chemical Engineering, vol. 55, no. 6, pp. 683–695, 1977.

72. B. N. Murthy, N. A. Deshmukh, A. W. Patwardhan, and J. B. Joshi, "Hollow self-inducing impellers: flow visualization and CFD simulation," Chemical Engineering Science, vol. 62, no. 14, pp. 3839–3848, 2007.

73. K. S. M. S. Raghava Rao, V. B. Rewatkar, and J. B. Joshi, "Critical impeller speed for solid suspension in mechanically agitated contactors," AIChE Journal, vol. 34, no. 8, pp. 1332–1340, 1988.

74. V. B. Rewatkar, K. S. M. S. Raghava Rao, and J. B. Joshi, "Critical impeller speed for solid suspension in mechanically agitated three-phase reactors. 1. Experimental part," Industrial and

Engineering Chemistry Research, vol. 30, no. 8, pp. 1770–1784, 1991.

75. B. N. Murthy, R. S. Ghadge, and J. B. Joshi, "CFD simulations of gas-liquid-solid stirred reactor: prediction of critical impeller speed for solid suspension," Chemical Engineering Science, vol. 62, no. 24, pp. 7184–7195, 2007.

76. J. B. Joshi and V. V. Ranade, "Computational fluid dynamics for designing process equipment: expectations, current status, and path forward," Industrial and Engineering Chemistry Research, vol. 42, no. 6, pp. 1115–1128, 2003.

77. D. Darmana, N. G. Deen, and J. A. M. Kuipers, "Detailed modeling of hydrodynamics, mass transfer and chemical reactions in a bubble column using a discrete bubble model," Chemical Engineering Science, vol. 60, no. 12, pp. 3383–3404, 2005.

78. D. Darmana, R. L. B. Henket, N. G. Deen, and J. A. M. Kuipers, "Detailed modelling of hydrodynamics, mass transfer and chemical reactions in a bubble column using a discrete bubble model: chemisorption of CO_2 into NaOH solution, numerical and experimental study," Chemical Engineering Science, vol. 62, no. 9, pp. 2556–2575, 2007.

79. D. Zhang, N. G. Deen, and J. A. M. Kuipers, "Euler-euler modeling of flow, mass transfer, and chemical reaction in a bubble column," Industrial and Engineering Chemistry Research, vol. 48, no. 1, pp. 47–57, 2009.

80. A. W. Vreman, "An eddy-viscosity sub-grid-scale model for turbulent shear flow: algebraic theory and applications," Physics of Fluids, vol. 16, no. 10, pp. 3670–3681, 2004.

81. K. Ekambara and J. B. Joshi, "Axial mixing in laminar pipe flows," Chemical Engineering Science, vol. 59, no. 18, pp. 3929–3944, 2004.

82. C. S. Mathpati, S. S. Deshpande, and J. B. Joshi, "Computational and experimental fluid dynamics of jet loop reactor," AIChE Journal, vol. 55, no. 10, pp. 2526–2544, 2009.

83. C. S. Mathpatii, M. V. Tabib, S. S. Deshpande, and J. B. Joshi, "Dynamics of flow structures and transport phenomena, 2. Relationship with design objectives and design optimization," Industrial and Engineering Chemistry Research, vol. 48, no. 17, pp. 8285–8311, 2009.

84. J. B. Joshi, V. S. Vitankar, A. A. Kulkarni, M. T. Dhotre, and K. Ekambara, "Coherent flow structures in bubble column reactors," Chemical Engineering Science, vol. 57, no. 16, pp. 3157–3183, 2002.

85. S. S. Deshpande, J. B. Joshi, V. R. Kumar, and B. D. Kulkarni, "Identification and characterization of flow structures in chemical process equipment using multiresolution techniques," Chemical Engineering Science, vol. 63, no. 21, pp. 5330–5346, 2008.

86. M. V. Tabib and J. B. Joshi, "Analysis of dominant flow structures and their flow dynamics in chemical process equipment using snapshot proper orthogonal decomposition technique," Chemical Engineering Science, vol. 63, no. 14, pp. 3695–3715, 2008.

87. M. V. Tabib, M. J. Sathe, S. S. Deshpande, and J. B. Joshi, "A hybridized snapshot proper orthogonal decomposition-discrete wavelet transform technique for the analysis of flow structures and their time evolution," Chemical Engineering Science, vol. 64, no. 21, pp. 4319–4340, 2009.

88. M. J. Sathe, I. H. Thaker, T. E. Strand, and J. B. Joshi, "Advanced PIV/LIF and shadowgraphy system to visualize flow structure in two-phase bubbly flows," Chemical Engineering Science, vol. 65, no. 8, pp. 2431–2442, 2010.

89. C. S. Mathpati, M. J. Sathe, and J. B. Joshi, "Reply to 'comments on dynamics of flow structures and transport phenomena—part I: experimental and numerical techniques for identification and energy content of flow structures'," Industrial and Engineering Chemistry Research, vol. 49, no. 9, pp. 4471–4473, 2010.

90. S. S. Deshpande, C. S. Mathpati, S. S. Gulawani, J. B. Joshi, and V. Ravi kumar, "Effect of flow structures on heat transfer in single and multiphase jet reactors," Industrial and Engineering Chemistry Research, vol. 48, no. 21, pp. 9428–9440, 2009.

91. S. S. Deshpande, M. V. Tabib, J. B. Joshi, V. Ravi Kumar, and B. D. Kulkarni, "Analysis of flow structures and energy spectra in chemical process equipment," Journal of Turbulence, vol. 11, article N5, pp. 1–39, 2010.

92. Z. Khan, C. S. Mathpati, and J. B. Joshi, "Comparison of turbulence models and dynamics of turbulence structures in bubble column reactors: effects of sparger design and superficial gas velocity," Chemical Engineering Science. In press.

Chapter 3

Corrosion and Inhibition Effects of Mild Steel in Hydrochloric Acid Solutions Containing Organophosphonic Acid

Manish Gupta[1], Jyotsna Mishra[1], and K. S. Pitre[2]

[1]Department of Chemistry, Scope College of Engineering, Bhopal, Madhya Pradesh 462047, India

[2]Department of Chemistry, Dr. H.S. Gour University, Sagar, Madhya Pradesh 470003, India

ABSTRACT

A study has been made on the mechanism of corrosion of mild steel and the effect of nitrilo trimethylene phosphonic (NTMP) acid as a corrosion inhibitor in acidic medium, that is, 10% HC1 using the weight loss method and electrochemical techniques, that is, potentiodynamic and galvanostatic polarization measurements. Although corrosion is a long-time process, but it takes place at a faster rate in the beginning which goes on decreasing with due course of time. The above-mentioned methods of corrosion rate determination furnish an average value for a long-time interval. Looking at the versatility and minimum detection limit of the voltammetric method, the authors have developed a new voltammetric method for the determination of corrosion rate at short-time intervals. The results of corrosion of mild steel in 10% HC1 solution with and without NTMP inhibitor at short-time intervals have been reported. The corrosion inhibition efficiency of NTMP is 93% after 24 h.

INTRODUCTION

Mild steel is a major material of construction. It is extensively used in chemical and allied industries for handling alkalis, acids, and salt solutions [1]. Hydrochloric (HC1) acid is the solvent most often employed for chemical cleaning. It attacks a wide range of scales [2]. Nitrogen compounds constitute the largest class of inhibitors for hydrochloric acid [3] solution. During the past decade a number of polymers and phosphonates have been used in different inhibitor compositions in aqueous and acid solutions [4]. They form stable complexes and some times act as detergent also. The role of inhibitor is the prevention of the adsorption of aggressive anions and reduction of the dissolution rate of the passivating oxide. In the present paper we have studied the action and effectiveness of nitrilo trimethylene phosphonic acid (NTMP) as inhibitor for corrosion of carbon steel in (10% HCl) acid solution.

Most of the methods are proposed in the field of corrosion rate determination, but they furnish an average value for a long-time interval [5]. Looking at the sensitivity and minimum detection limits [6, 7] of polarographic techniques, that is, direct current polarography (DCP), differential pulse polarography (DPP), and voltammetric, that is, differential pulse anodic stripping voltammetric (DPASV) method has been used to determine the corrosion rates and inhibition efficiency of NTMP of mild steel in 10% HC1 solution at short time intervals. Significantly it has also been possible to determine the corrosion rates simultaneously with respect to Fe(II) and Fe(III) which is not possible using other methods prevalent in the field. Their results have been discussed in the paper.

EXPERIMENTAL

Chemicals and Reagents

All the chemicals used were of anal R/BDH grade. The inhibitor (NTMP, 99.9% pure) was synthesized at the Central Research Institute For Chemistry, Budapest, Hungery. Experiments were carried out in 10% HC1 solution. Mild Steel specimens 50 * 20 * 4 mm size having composition (C-0.23%, P-0.05%, S-0.055%) were used in experiments. These specimen were polished following the usual procedure [8]. All the measurements were carried out at room temperature 30°C.

Gravimetric Measurements

After abrasion with 600 grit papers all samples were degreased with acetone, pickled in 15% HC1 solution, washed under running tap water, rinsed in distilled water, dried with acetone, and weighed. The variation of solution corrosiveness and corrosion rate as a function of time was investigated using the planned interval test technique [8]. Three weighed specimens were introduced into the solution. These were removed after various exposure times, giving damage factors referred to as A_1 for 22 h, A_t for 3 h, and A_t+1 for 25 h. A fourth specimen (B) was introduced into the cell for the last period for 3 h. The values of A_1, B (measured values) and A_c (calculated from A_1, and A_{t+1}) relate to the same testing period, but under different conditions. This technique is schematically illustrated in Figure 1.

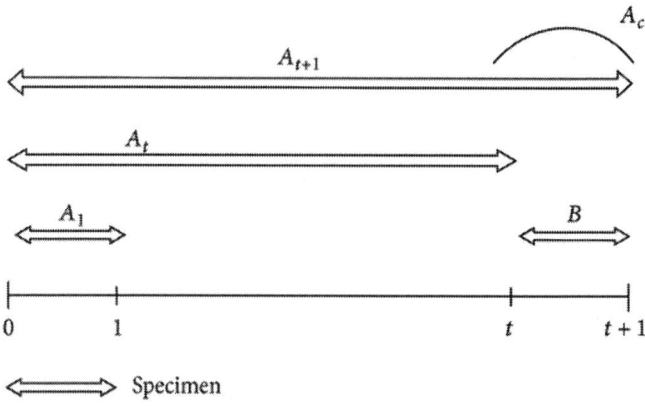

Specimen

Figure 1: Procedure of "planned interval test" technique identical specimen all placed in the some corrosive fluid imposed conditions of the test kept constant for entire time A_{t+1}. The terms A_l, A_t, A_{t+1}, and B represent corrosive damage experienced by each specimen, Ac is calculated by substracting A_t from A_{t+1}.

After each experiment, the specimens were cleaned by washing in water and acetone then weighed; from these measurements the corrosion rates were calculated. The methods of evaluating the results and combinations of situations are summarized in Table 1.

Table 1: Results of planned interval test (PIT) on mild steel in 10% HCl solution

S. no.	Period	Without inhibitor		With inhibitor		% inhibition
		Concentration (milligram × 102)	Corrosion rate (milligram centimeter – 2 hour – 1 × 102)	Concentration (milligram × 102)	Corrosion rate (milligram centimeter – 2 hour–1 × 102)	
1	1 (3 h)	18.96	21.2	5.21	5.82	72
2	t (22 h)	31.05	4.70	6.32	0.96	79
3	t+1 (22 + 3 h)	34.58	4.64	6.40	0.85	82
4	B (3 h)	3.33	3.72	0.50	0.50	96
5	C (3 h)	3.53	3.94	0.08	0.08	97

Potentiodynamic Polarization Measurement

The specimen was molded into epoxy resin in order to cover the sides of the rod, while the circular cross section area of the cylinder was exposed to the solution. The specimen was polished and washed with distilled water before the experiment. The potential scan path was 50 mV/5 min. The observed current was plotted against applied potential.

Galvanostatic Polarization Measurement

The galvanostatic polarization measurements were also made in 10% HC1 solution on 10 cm² cylindrical surface of the working electrode. Saturated calomel electrode (SCE) was used as a reference electrode. An AJCO Electronics (model VT S5016) attached to a multiplex galvanometer was used for the galvanostatic studies. The polarization currents were measured by using a cycle of 10 s and changing potential by 10 mV. Both anodic and cathodic cycles were used alternately. The corrosion current density was plotted against the applied potential.

Voltammetric and Polarographic Measurements

An Elico Pulse Polarograph (model CL-90) was used for these studies. A cell consisting of three electrodes, namely, saturated calomel electrode as a reference, a coiled platinum wire as an auxiliary, and a dropping mercury electrode (for DCP and DPP)/glassy carbon fiber electrode (for DPASV) as a working electrode, was used. The test specimens were polished as discussed earlier, and one such specimen was suspended in 10% HCl solution at room temperature (25°C). Nitrogen gas was bubbled throughout the solution through the experiment to avoid the oxidation of dissolved Fe(II) to Fe(III). A definite aliquot of the solution was withdrawn from the test solution at different intervals (5, 10, 20 min, 1, 2, and 24 h), and polarograms and voltammograms were recorded in deaerated 0.1 M ammonium tartrate + 0.001% gelatin at pH 9.0 ± 0.1. The pH of the test solution was adjusted using ammonia solution. A similar experiment was performed using 170 mg/L NTMP in 10% HC1 solution.

RESULTS AND DISCUSSION

Gravimetric Measurement

The results of weight loss determination by gravimetric are shown as corrosion rates in the presence and absence of NTMP inhibitor in Table 1. The data obtained for the periods A_1 and B suggest that solution corrosiveness increases in the base solution (10% HCl) while in the presence of NTMP it decreases ($B<A_1$). The higher corrosivity of the solution can be attributed to the increase in concentration of iron ions in the solution. This suggests that NTMP molecules reduce the stimulating effect of iron ions. By comparison with the data relating to B and A_c it can be concluded that metal corrodibility decreases ($Ac < B$) as a function of time in the presence of the NTMP. This indicates that the formation of an inhibitor film requires some time; that is, the inhibition process has an induction period. The increased inhibition effect of NTMP may be explained on the argument that nitrogen of NTMP facilitates complex formation between metal ions coming out in the solution with NTMP and also the formation of complexed film at the metal surface.

Potentiodynamic Polarization Measurements

Figure 2 shows the results of potentiodynamic polarization measurements in the presence and absence of NTMP inhibitor in 10% HC1 solution. The curve clearly shows the cathodic and anodic inhibition properties of inhibitor. It is quite clear from the curves that the corrosion potentials are more positive (anodic) than the rest of potentials. The inhibitor shows a remarked effect in lowering the current. However, in the cathodic region the effect is not significant. Thus it could be conduced that NTMP affects the anodic part of the corrosion process. Thus they work as anodic protector.

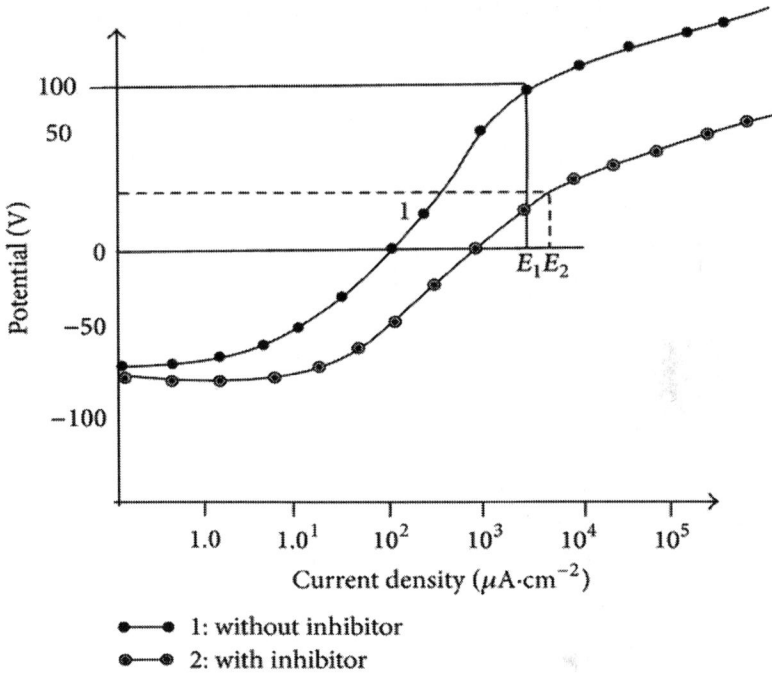

Figure 2: Potentiodynamic polarization curves of carbon steel in 10% HC1 with, and without inhibitor.

Galvanostatic Polarization Measurement

Figure 3 depicts the galvanostatic polarization curves for the corrosion of mild steel in 10% HC1 solution with and without NTMP inhibitor. It is clear from the figure that both anodic and cathodic currents decrease in the presence of inhibitor. However, the corrosion potentials are shifted to more electropositive values with NTMP inhibitor, but the Tafel slopes of the curves remain almost the same. The shift in corrosion potential with NTMP is +50 mV. Thus, it could be concluded that the film formed by the inhibitor acts through the blocking effect, mainly on the anodic reaction.

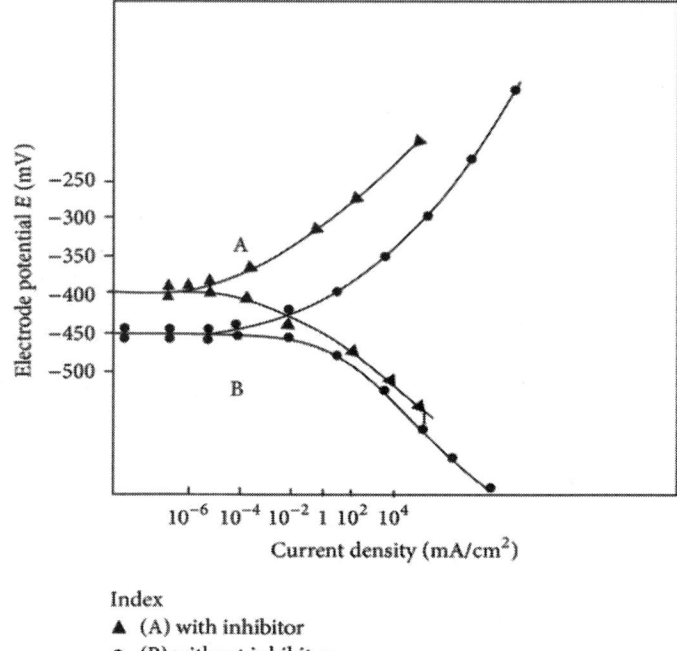

Index
▲ (A) with inhibitor
● (B) without inhibitor

Figure 3: Galvanostatic polarization curves of carbon steel in 10% HC1 with and without inhibitor.

Voltammetric and Polarographic Measurements

Figures 4(a), 4(b), and 4(c) are the direct current polarogram (DCP), differential pulse polarogram (DPP), and differential pulse anodic stripping voltammogram (DPASV) for corrosion sample after 5 min in 0.1 M ammonium tartrate and 0.001% gelatin at pH 9.0 ± 0.1. The half-wave potential ($E_{1/2}$)/peak potential (E_p) values for Fe(II) and Fe(III) are −0.81 V/−0.83 V and −1.23/−1.25 V, respectively, in DCP/DPP mode, and E_p values are −0.4 V and −0.72 V for Fe(II) and Fe(III) in DPASV mode [9]. Figure 5 shows the DPASV curves of corrosion mixtures of mild steel at different time intervals. These curves show that by using voltammetric and polarographic methods, it is possible to analyze the presence of Fe(II) and Fe(III) in the solution, simultaneously. Fe(II) and

Fe(III) produce two separate well-defined polarographic peaks in each case. The wave height/peak height for Fe(II) and Fe(III) over the whole of the working concentration range in DCP, DPP, and DPASV modes is proportional to the Fe(II) and Fe(III) concentrations. The corrosion rates at different time intervals with respect to Fe(II) and Fe(III) are summarized in Table 2.

Table 2: Corrosion rates with respect to concentration of Fe(II), Fe(III), and total Fe(II + III) for mild steel in 10% HCl without inhibitor using voltammetric and polarographic methods

(a)DPP mode						
Time	**Fe(II)**		**Fe(III)**		**Total Fe(II + III)**	
	Conc*	CR**	Conc*	CR**	Conc*	CR**
(1) 5 min	5.91	2.37	8.23	3.31	14.14	5.69
(2) 10 min	6.13	1.23	9.89	1.99	16.02	3.22
(3) 20 min	7.93	0.79	11.21	1.12	19.14	1.92
(4) 1 h	12.16	0.40	12.55	0.42	24.17	0.83
(5) 2 h	15.98	0.26	16.81	0.28	32.78	0.55
(6) 24 h	50.57	0.07	56.17	0.07	107.28	0.15

(b)DPASV mode						
Time	**Fe(II)**		**Fe(III)**		**Total Fe(II + III)**	
	Conc*	CR**	Conc*	CR**	Conc*	CR**
(1) 5 min	5.92	2.38	8.24	3.31	14.16	5.70
(2) 10 min	6.15	1.23	9.87	1.98	16.02	3.22
(3) 20 min	7.93	0.79	11.21	1.12	19.11	1.92
(4) 1 h	12.16	0.40	12.50	0.42	24.57	0.82
(5) 2 h	15.98	0.26	16.80	0.28	32.78	0.55
(6) 24 h	50.57	0.07	56.17	0.07	107.28	0.15

*Corrosion concentration $\times\ 10^2$ milligram.

**Corrosion rate $\times\ 10^2$ milligram centimeter^{-2} hour^{-1}.

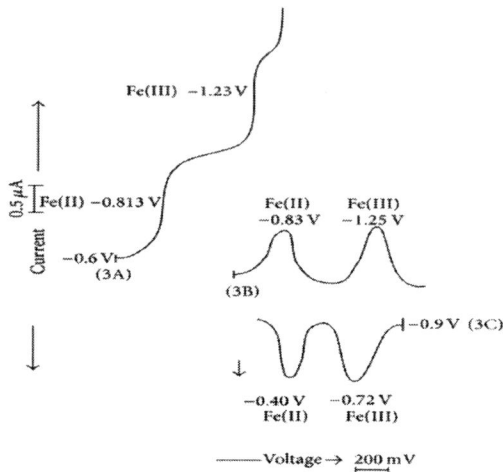

Figure 4: (a) Direct current polarogram, (b) Differential pulse polarogram, (c) Differential pulse anodic stripping voltammogram of corrosion solution (10 mL) in 0.1 M ammonium tartrate + 0.001% gelatin at pH 9.0 ± 0.1 after 5 min. of exposure.

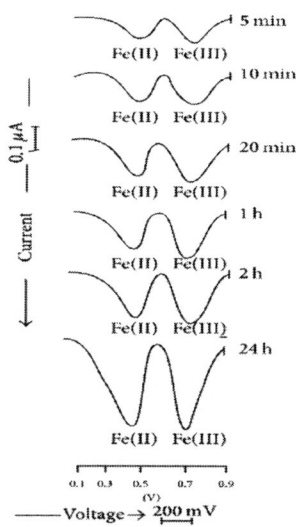

Figure 5: Differential pulse anodic stripping voltammogram of the corrosion solution of carbon steel in 10% HC1 at different time intervals in ammonium tartrate + 0.01% gelatin at pH 9.0 ± 0.1.

Figure 6 shows the DPASV curves for the corrosion solution to which NTMP (170 mg/L) has been added. A shift in the E_p value for each of the Fe(II) and Fe(III) indicates the M: NTMP complex formation in the solution.

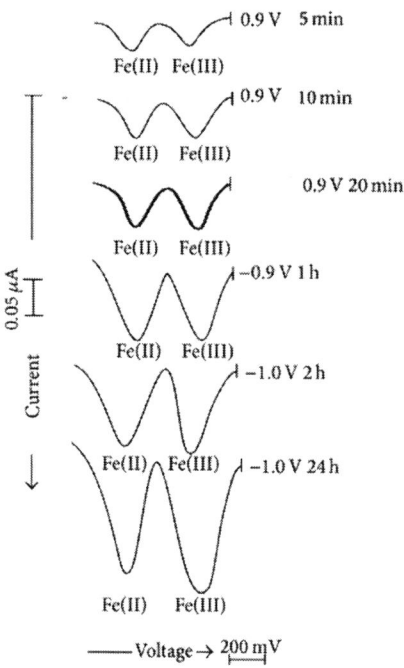

Figure 6: Differential pulse anodic stripping voltammogram of the corrosion solution of carbon steel, 0.1 M ammonium tartrate + 0.001% gelatin at pH 9.0 ± 0.1 at different time intervals in presence of Inhibitor.

Since the shift in $E_{1/2}$ and E_p values of Fe(III) is more than that of Fe(II) in complexed system, it could be concluded that Fe(III) is more reactive in complex formation with NTMP than Fe(II) [10]. The corrosion rates in the presence of NTMP inhibitor have been tabulated in Table 3. A perusal of the data in Tables 1, 2, and 3clearly shows that the rate of corrosion is very fast in the beginning of the experiment, which goes on decreasing with due course of time. The corrosion inhibition efficiency of NTMP is 55%, 69%, 75%, 81%, and 84% after 5, 10, 20 min, 1, and 2 h, and it becomes 93% after 24 h, respectively.

Table 3: Corrosion rates with respect to Fe (II) + Fe (III) for mild steel in 10% HCl solution with NTMP inhibitor (170 mg/L)

Time	DPASV mode			DPP mode		
	Conc*	CR**	% inhibition	Conc*	CR**	% inibition
(1) 5 min	6.28	2.530	55.6	6.26	2.52	55.7
(2) 10 min	6.41	1.290	59.9	6.41	1.290	59.9
(3) 20 min	6.99	0.693	63.9	6.99	0.693	63.9
(4) 1 h	7.18	0.235	71.6	7.02	0.235	71.6
(5) 2 h	7.21	0.119	80.0	7.12	0.119	80.6
(6) 24 h	7.91	0.010	93.3	7.81	0.010	93.3

*Corrosion concentration $\times$ 102 milligram.

**Corrosion rate $\times 10^2$ milligram centimeter^{-2} hour^{-1}.

Thus, the inhibitive efficiency of NTMP (organophosphonic acid) will depend on the structure of the protective inhibitor film formed by the dissolved Fe(II) and Fe(III) species on the metal surface. The NTMP used as an inhibitor for the present study has the following structure:

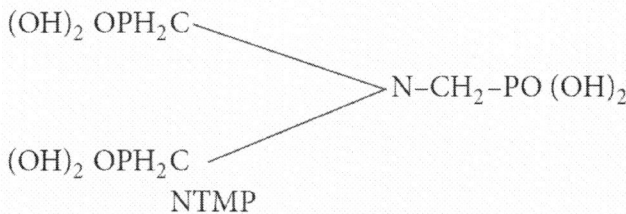

$$\text{(OH)}_2 \text{ OPH}_2\text{C} \diagdown$$
$$\text{N-CH}_2\text{-PO (OH)}_2$$
$$\text{(OH)}_2 \text{ OPH}_2\text{C} \diagup$$
NTMP

The nitrogen atom acts as the reaction centre in the complexation reaction with the metal ions.

A comparison of the corrosion rate data as calculated by gravimetric, galvanostatic, potentiodynamic polarization, and voltammetric methods clearly speaks that NTMP is a good inhibitor for corrosion inhibition of mild steel in acidic medium, that is, 10% HC1.

From the above data and ongoing discussion, it could be concluded that polarographic and voltammetric methods give dependable results for corrosion rates with respect to each species appearing in the solution, that is, Fe(II) and Fe(III). It is also clear that the proposed polarographic/voltammetric methods are highly sensitive for corrosion rate determination at short-time intervals and that too with respect to each species present in the solution, which is otherwise not possible using methods prevalent in the field of corrosion rate determination [11, 12] and also in the field of trace analysis, namely, atomic absorption spectrometry (AAS).

ACKNOWLEDGMENTS

The authors are thankful to the M. P. Council of Science and Technology (MPCST) Bhopal, for providing research fellowship.

REFERENCES

1. A. Singh, E. E. Ebenso, M. A. Quraishi, et al., "Corrosion inhibition of carbon steel in HCl solution by some plant extracts," International Journal of Corrosion, vol. 2012, Article ID 897430, 20 pages, 2012.

2. T. Horvath, E. Kalman, G. Kutsan, and A. Rauscher, "Corrosion of mild steel in hydrochloric acid solutions containing organophosphonic acids," British Corrosion Journal, vol. 29, no. 3, pp. 215–218, 1994.

3. P. Mohan, R. Usha, G. P. Kalaighan, and V. S. Murlidharan, "Inhibition effect of benzohydrazide derivatives on corrosion behaviour of mild steel in 1 M HCl," Journal of Chemistry, vol. 2013, Article ID 541691, 7 pages, 2013.

4. H. Ryu, N. Sheng, T. Outsuka, S. Fugita, and H. Kajtyama, "Polypyrrole film on 55% al–zn-coated steel for corrosion prevention," Corrosion Science, vol. 56, pp. 67–77, 2012.

5. A. J. Freelman, Materials Performance, vol. 23, pp. 9–11, 1984.

6. T. E. Edmonds and J. I. Guoliang, "Carbon fibre micro-electrodes in the differential pulse voltammetry of copper ions," Analytica Chimica Acta, vol. 151, pp. 99–108, 1983.

7. A. Wachter and R. S. Treseder, "Corrosion testing evaluation of metals for process equipment,"Chemical Engineering Progress, vol. 43, pp. 315–326, 1947.

8. E. Kálmán, F. H. Kármán, J. Telegdi, B. Várhegyi, J. Balla, and T. Kiss, "Inhibition efficiency of n-containing carboxylic and carboxy-phosphonic acids," Corrosion Science, vol. 35, pp. 1477–1481, 1993

9. V. K. Chitale and K. S. Pitre, Reviews in Analytical Chemistry, vol. 6, pp. 177–184, 1982.

10. V. Rai and K. S. Pitre, "Corrosion behaviour of carbon steel in DTPMP inhibited neutral medium,"Indian Journal of Chemistry A, vol. 42, no. 1, pp. 106–108, 2003.

11. J. Shukla and K. S. Pitre, "Corrosion and inhibition kinetics of PVA polymer on carbon steel in sulfuric acid solution," Indian Journal of Chemistry A, vol. 44, no. 11, pp. 2270–2273, 2005.

12. J. Shukla, P. Jain, and K. S. Pitre, "Inhibitive action of thiourea plus Ca towards corrosion of brass in acidic solution," Corrosion Reviews, vol. 22, pp. 145–156, 2004.

4

Modeling and Control of Distillation Column in a Petroleum Process

Vu Trieu Minh and Ahmad Majdi Abdul Rani

Mechanical Engineering Department, Universiti Teknologi PETRONAS, Bandar Seri Iskandar, 31750 Tronoh, Perak Darul Ridzuan, Malaysia

ABSTRACT

This paper introduces a calculation procedure for modeling and control simulation of a condensate distillation column based on the energy balance (L-V) structure. In this control, the reflux rate L and the boilup rate V are used as the inputs to control the outputs of the purity of the distillate overhead and the impurity of the bottom products. The modeling simulation is important for process dynamic analysis and the plant initial design. In this paper, the modeling and simulation are accomplished over three phases: the basic nonlinear model of the plant, the full-order linearised model, and the reduced-order linear model. The reduced-order linear model is then used as the reference model for a model-reference adaptive control (MRAC) system to verify the applicable ability of a conventional adaptive controller for a distillation column dealing with the disturbance and the model-plant mismatch as the influence of the plant feed disturbances.

INTRODUCTION

Distillation is the most popular and important separation method in the petroleum industries for purification of final products. Distillation columns are made up of several components, each of which is used either to transfer heat energy or to enhance mass transfer. A typical distillation column contains a vertical column where trays or plates are used to enhance the component separations, a reboiler to provide heat for the necessary vaporization from the bottom of the column, a condenser to cool and condense the vapor from the top of the column, and a reflux drum to hold the condensed vapor so that liquid reflux can be recycled back from the top of the column.

Calculation of the distillation column in this paper is based on a real petroleum project to build a gas processing plant to raise the utility value of condensate. The nominal capacity of the plant is 130 000 tons of raw condensate per year based on 24 operating hours per day and 350 working days per year. The quality of the output products is the purity of the distillate,x_D, higher than or equal to 98% and the impurity of the bottoms,x_B, less/equal than 2%. The basic feed stock data and its actual compositions are based on [1].

Most of distillation control systems, either conventional or advanced, assume that the column operates at a constant pressure. Pressure fluctuations make the control more difficult and reduce the performance. The L-Vstructure, which is called energy balance structure, can be considered as the standard control structure for a dual composition control distillation. In this control structure the liquid flow rate L and the vapor flow rate V are the control inputs. The objective of the controller is to maintain the product outputs concentrations F and despite the disturbance in the feed flow and the feed concentration c_F (Figure 1).

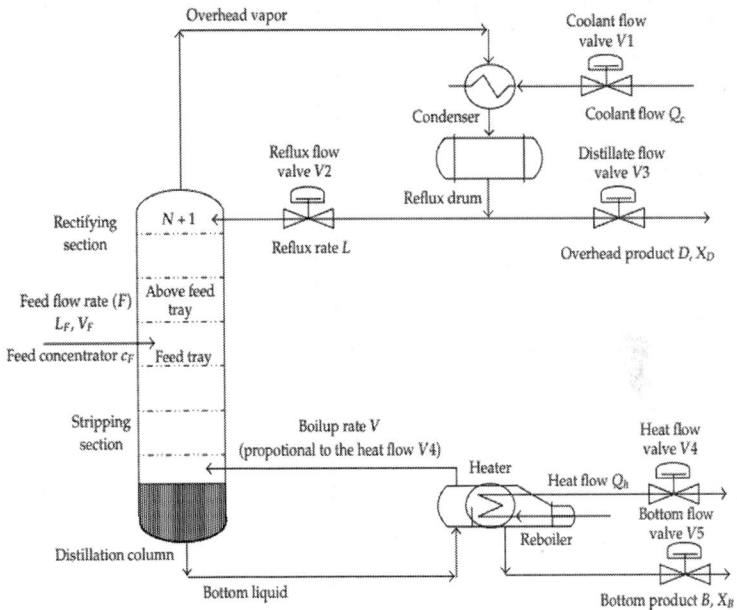

Figure 1: Distillation flowsheet.

The goals of this paper are twofold: first, to present a theoretical calculation procedure of a condensate column for simulation and analysis as an initial step of a project feasibility study, and second, for the controller design: a reduced-order linear model is derived such that it best reflects the dynamics of the distillation process and used as the reference model for a model-reference adaptive control (MRAC) system to verify the ability of a conventional adaptive controller for a distillation process dealing with the disturbance and the plant-model mismatch as the influence of the feed disturbances.

In this study, the system identification is not employed since experiments requiring a real distillation column are still not implemented yet. So that a process model based on experimentation on a real process cannot be done. A mathematical modeling based on physical laws is performed instead. Further, the MRAC controller model is not suitable for handling the process constraints on inputs and outputs as shown in [2] for a coordinator model predictive control (MPC). In this paper, the calculations and simulations are implemented by using MATLAB (version 7.0) software package.

PROCESS MODEL AND SIMULATION

The feed can be considered as a pseudobinary mixture of Ligas (iso-butane, n-butane and propane) and Naphthas (iso-pentane, n-pentane, and higher components). The column is designed with N=14 trays. The model is simplified by lumping some components together (pseudocomponents) and modeling of the column dynamics is based on these pseudocomponents only [3].

For the feed section, the operating pressure at the feed section is given at 4.6 atm. The feed temperature for the preheater is the temperature at which the required phase equilibrium is established. Consulting the equilibrium flash vaporization (EFV) curve at 4.6 atm, the required feed temperature is selected at corresponding to the point of 42% of the vapor phase feed rate V_F.

For the rectifying section, the typical pressure drop per tray is 6.75 kPa. Thus, the pressure at the top section is 4 atm. Also consulting the Cox chart, the top section temperature is determined at 46°C Then, we can calculate the reflux flow rate L via the energy balance equation.

For the stripping section, the column base pressure is approximately the pressure of the feed section (4.6 atm) because the pressure drop across this section is neglected. Consulting the EFV curve and the Cox chart, the equilibrium temperature at this section (4.6 atm) is determined at 144°C Then, we can calculate the reboiler duty or the heat input Q_B to increase the temperature of stripping section from 118°C to 144°C.

Table 1 summarizes the initial calculated data for the main streams of input feed flow rate (Condensate), output distillate overhead product: (LPG) and output bottom product (Raw gasoline).

Table 1: The main streams

Stream	Condensate	LPG	Raw gasoline
Temperature (°C)	118	46	144
Pressure (atm)	4.6	4.0	4.6
Density (kg/m³)	670	585	727
Volume flow rate (m³/h)	22.76	8.78	21.88
Mass flow rate (kg/h)	15480	5061	10405
Plant capacity (ton/year)	130000	43000	87000

The vapor boilup V generated by the heat input to the reboiler is calculated as [4]: $V = (Q_B - Bc_B(t_B - t_F))/\lambda$ (kmole/h), where Q_B is the heat input (kJ/h); B is the flow rate of bottom product (kg/h); c_B is the specific heat capacity (kJ/kg· °C); t_F is the inlet temperature(°C); t_B is the outlet temperature (°C); λ is the latent heat or the heat of vaporization (kJ/kg).The latent heat at any temperature is described in terms of the latent heat at the normal boilingpoint [5] $\lambda = \gamma\lambda_B(T/T_B)$, where λ is the latent heat at the absolute temperature T in degrees Rankine (°R); λ_B is the latent heat at the absolute normal boiling point T_B in degrees Rankine (°R); and γ is the correction factor obtained from the empirical chart.

Major design parameters to determine the liquid holdup on tray, column base and reflux drum are calculated mainly based on [6–8].

Velocity of vapor phase is arising in the column

$\omega_n = C\sqrt{(\rho L - \rho G)/\rho G}$ (m/s), where ρL (kg/m³) is the density of liquid phase; ρG (kg/m³) is the density of vapor phase; C is the correction factor depending flow rates of two-phase flows.

The actual velocity ω is normally selected at $\omega = (0.80 - 0.85)\omega_n$ for paraffinic vapor.

The diameter of the column is calculated on the formula: $D_k = \sqrt{4V_m / 3600\pi\omega}$ (m), where V_m (kmole/h) is the mean flow of vapor in the column.

The holdup in the column base is $M_B = (\pi H_{NB}D^2_k/4) (\rho_B/(MW)_B)$ (kmole), where H_{NB} (m) is the normal liquid level in the column base; $(MW)_B$ is the molar weight of the bottom product (kg/kmole); ρ_B is the density of the bottom product (kg/m³).

Similarly, the holdup on each tray is $M = (0.95\pi h_T D^2_k/4(\rho_T/(MW)_T)$ (kmole), where h_T is the average depth of clear liquid on a tray (m); $(MW)_T$ is the molar weight of the liquid holdup on a tray (kg/kmole); ρ_T is the mean density of the liquid holdup on a tray (kg/m3).And the holdup in the reflux drum $M_D = 5(L_f + V_f)/60$ (kmole), where L_f is the reflux flowrate (kmole/h); V_f is the distillate flow rate (kmole/h).

The rate of accumulation of material in a system is equal to the amount entered and generated, less the amount leaving and consumed within the system. The model is simplified under assumptions in [9].

- Constant relative volatility throughout the column and the vapor-liquid equilibrium relation can be expressed by

$$y_n = \frac{\alpha x_n}{1 + (\alpha - 1)x_n},$$

(2.1)

where x_n is the liquid concentration on nth stage; y_n is the vapor concentration on nth stage; α is the relative volatility.

- The overhead vapor is totally condensed.
- The liquid holdups on each tray, the condenser, and the reboiler are constant and perfectly mixed.

The holdup of vapor is negligible throughout the system

- The molar flow rates of the vapor and liquid through the stripping and rectifying sections are constant.

Under these assumptions, the dynamic model can be expressed by the following equations:

- condenser (n=N+2):

$$M_D \dot{x}_n = (V + V_F)y_{n-1} - Lx_n - Dx_n,$$

(2.2)

- tray n(n=f+2 to N+1):

$$M\dot{x}_n = (V + V_F)(y_{n-1} - y_n) + L(x_{n+1} - x_n),$$

(2.3)

- tray above the feed flow (n=f+1):

$$M\dot{x}_n = V(y_{n-1} - y_n) + L(x_{n+1} - x_n) + V_F(y_F - y_n),$$

(2.4)

- tray below the feed flow (n=f):

$$M\dot{x}_n = V(y_{n-1} - y_n) + L(x_{n+1} - x_n) + L_F(x_F - x_n),$$

(2.5)

- tray n(n=2 to f-1):

$$M\dot{x}_n = V(y_{n-1} - y_n) + (L + L_F)(x_{n+1} - x_n),$$

(2.6)

- reboiler (n=1):

$$M_B\dot{x}_1 = (L + L_F)x_2 - Vy_1 - Bx_1.$$

(2.7)

Although the model is simplified, the representation of the distillation system is still nonlinear due to the vapor-liquid equilibrium relationship between y_n and x_n in (2.1).

The distillation process simulation is done using Matlab Simulink as shown in Figure 2. The dynamic model is represented by a set of 16 nonlinear differential equations: $x_1 = x_2$ is the liquid concentration in bottom; x_2 is the liquid concentration in the 1st tray, x_3 is the liquid concentration in the 2nd tray; ...;x_{15} is the liquid concentration in the 14th tray; and $x_{16}=x_D$ is the liquid concentration in the distillate.

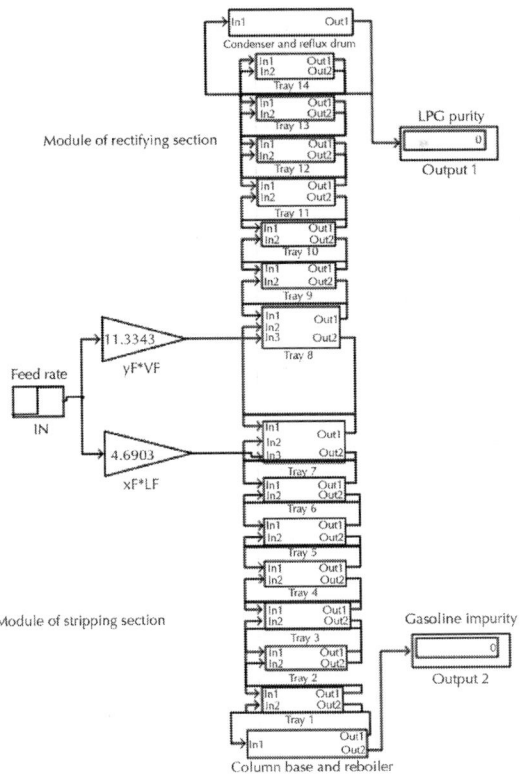

Figure 2: Model simulation with Matlab Simulink.

If there are no disturbance in the operating conditions as shown in Figure 3, the system is to reach the steady state such that the purity of the distillate product x_D equals 0.9654 and the impurity of the bottoms product equals 0.0375.

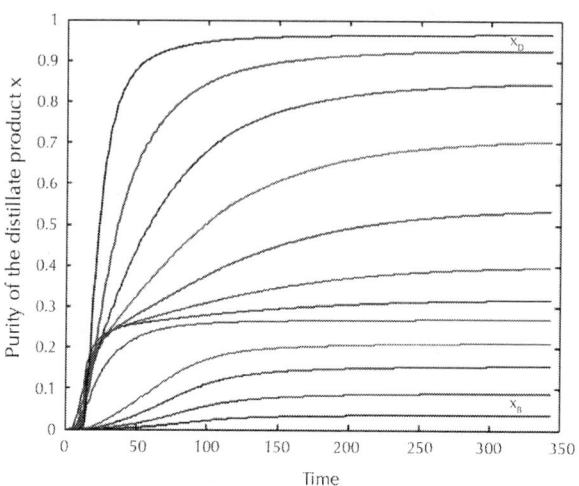

Figure 3: The steady-state values of concentrations x_n on each tray.

Table 2 indicates the steady-state values of concentration of x_n and y_n on each tray.

Table 2: The steady state values of concentrations x_n and y_n on each tray

Stage	Bottom	Tray 1	Tray 2	Tray 3	Tray 4	Tray 5	Tray 6	Tray 7
x_n	0.0375	0.0920	0.1559	0.2120	0.2461	0.2628	0.2701	0.2731
y_n	0.1812	0.3653	0.5120	0.6044	0.6496	0.6694	0.6776	0.6809
Stage	Tray 8	Tray 9	Tray 10	Tray 11	Tray 12	Tray 13	Tray 14	Distillate
x_n	0.2811	0.3177	0.3963	0.5336	0.7041	0.8449	0.9369	0.9654
y_n	0.6895	0.7256	0.7885	0.8666	0.9311	0.9687	0.9883	0.9937

Since the feed stream depends on the upstream processes, the changes of the feed stream can be considered as disturbances including the changing in feed flow rates and feed compositions. Simulations with these disturbances indicate that the quality of the output products gets worse if the disturbances exceed some certain ranges as shown in Table 3.

Table 3: Product quality depending on the change of the feed rates

	Purity of the distillate product $x_D(\%)$	Impurity of the bottoms product $x_B(\%)$
Normal feed rate	96.54	3.75
Reduced feed rate 10%	90.23	0.66
Increased feed rate 10%	97.30	11.66

The designed system does not achieve the operational objective of the product quality ($x_D \geq 0.98$ and $x_B \leq 0.02$) and the product quality will get worse dealing with disturbances. Hence we will use an adaptive controller—MRAC—to take the system from these steady-state outputs of $x_D = 0.9654$ and $x_B = 0.0375$ to the desired output targets.

LINEARIZATION OF THE DISTILLATION PROCESS

In order to obtain a linear control model for this nonlinear system, we assume that the variables deviate only slightly from some operating conditions [10]. Then the nonlinear equation in (2.1) can be expanded into a Taylor's series. If the variation $x_n - x_n$ is small, we can neglect the higher-order terms in $x_n - x_n$. The linearization of the distillation column leads to a 16th order linear model in the state space form:

$$\dot{z}(t) = Az(t) + Bu(t),$$

$$y(t) = Cz(t), \tag{3.1}$$

where

$$z(t) = \begin{bmatrix} x_1(t) - \overline{x}_{1\ \text{Steady State}} \\ x_2(t) - \overline{x}_{2\ \text{Steady State}} \\ \vdots \\ x_{16}(t) - \overline{x}_{16\ \text{Steady State}} \end{bmatrix}, \quad u(t) = \begin{bmatrix} L(t) - \overline{L}_{\text{Steady State}} \\ V(t) - \overline{V}_{\text{Steady State}} \end{bmatrix},$$

$$y(t) = \begin{bmatrix} x_1(t) - \overline{x}_{1\ \text{Steady State}} \\ x_{16}(t) - \overline{x}_{16\ \text{Steady State}} \end{bmatrix}.$$

(3.2)

The matrix A elements (n for each stage) are

- reboiler:

$$a_{1,1} = -\frac{\left(K_1 \overline{V} + B \right)}{M_B}, \quad a_{1,2} = \frac{\left(\overline{L} + \overline{L}_F \right)}{M_B},$$

(3.3)

- stripping section, tray 1 ÷ 6:

$$a_{n,n-1} = \frac{\left(K_{n-1} \overline{V} \right)}{M}, \quad a_{n,n} = -\frac{\left(K_n \overline{V} + \overline{L} + L_F \right)}{M}, \quad a_{n,n+1} = \frac{\left(\overline{L} + L_F \right)}{M},$$

(3.4)

- feeding section, tray 7 ÷ 8:

For n=8,

$$a_{8,7} = \frac{\left(K_7 \overline{V} \right)}{M}, \quad a_{8,8} = -\frac{\left(K_8 \overline{V} + \overline{L} + L_F \right)}{M}, \quad a_{8,9} = \frac{\left(\overline{L} \right)}{M},$$

$$a_{9,8} = \frac{\left(K_8 \overline{V} \right)}{M}, \quad a_{9,9} = -\frac{\left(K_9 \overline{V} + \overline{L} \right)}{M}, \quad a_{9,10} = \frac{\left(\overline{L} \right)}{M},$$

(3.5)

rectifying section, tray 9 ÷ 14:

$$a_{n,n-1} = \frac{\left(K_{n-1} \left(\overline{V} + V_F \right) \right)}{M}, \quad a_{n,n} = -\frac{\left(K_n \left(\overline{V} + V_F \right) + \overline{L} \right)}{M},$$

$$a_{n,n+1} = \frac{\left(\overline{L} \right)}{M}$$

(3.6)

- condenser:

$$_{16,15} = \frac{\left(K_{15} \left(\overline{V} + V_F \right) \right)}{M_D}, \quad a_{16,16} = -\frac{\left(\overline{L} + D \right)}{M_D}$$

(3.7)

where K_n is the linearized Vapor-Liquid Equilibria (VLE) constant:

$$K_n = \frac{dy_n}{dx_n} = \frac{\alpha}{(1 + (\alpha - 1)x_n)^2} = \frac{5.68}{(1 + 4.68x_n)^2}.$$

(3.8)

The matrix B elements are

$$\text{for } n = 1, \quad b_{1,1} = \frac{(\overline{x}_2)}{M_B}L, \quad b_{1,2} = -\frac{(\overline{y}_1)}{M_B}V,$$

$$\text{for } n = 2 \div 15, \quad b_{n,1} = \frac{(\overline{x}_{n+1} - \overline{x}_n)}{M}L, \quad b_{n,2} = -\frac{(\overline{y}_n - \overline{y}_{n-1})}{M}V,$$

$$\text{for } n = 16, \quad b_{16,1} = -\frac{(\overline{x}_{16})}{M_D}L, \quad b_{16,2}\frac{(\overline{y}_{15})}{M_D}V.$$

(3.9)

The output matrix C is

$$C = \begin{vmatrix} 1 & 0 & 0 & 0 & 0 & 0 & 0 & 0 & 0 & 0 & 0 & 0 & 0 & 0 & 0 & 0 \\ 0 & 0 & 0 & 0 & 0 & 0 & 0 & 0 & 0 & 0 & 0 & 0 & 0 & 0 & 0 & 1 \end{vmatrix}.$$

(3.10)

The full-order linear model which represents a two inputs-two outputs plant in equation in (3.3) can be expressed as a reduced order linear model as in [11, 12]:

$$\begin{bmatrix} x_D \\ x_B \end{bmatrix} = \frac{1}{1 + \tau_c s}G(0)\begin{bmatrix} L \\ V \end{bmatrix},$$

(3.11)

where $G(0)$ is the steady-state gain: $G(0) = (CA^{-1}B, \tau_c$ is the time constant:

$$\tau_c = \frac{M_I}{I_s \ln S} + \frac{M_D(1 - x_D)x_D}{I_s} + \frac{M_B(1 - x_B)x_B}{I_s},$$

(3.12)

whereM_I (kmole) is the total holdup of liquid inside the column;M_D (kmole) is the liquid

holdup in the condenser;M_B (kmole) is the liquid holdup in the reboiler; I_s is the "impurity

sum"; S is the separation factor.

As the result of calculation, the reduced-order linear model of the plant is a first-order system with a time constant of $\tau_c = 1.9588(h)$:

$$\begin{bmatrix} x_D \\ x_B \end{bmatrix} = \frac{1}{1 + 1.9588s} \begin{bmatrix} 0.0042 & -0.0062 \\ -0.0052 & 0.0072 \end{bmatrix} \begin{bmatrix} L \\ V \end{bmatrix}.$$

(3.13)

Equation (3.13) is equivalent to the following linear model in state space:

$$\dot{z}_r(t) = \begin{vmatrix} -0.5105 & 0 \\ 0 & -0.5105 \end{vmatrix} z_r(t) + \begin{vmatrix} 1 & 0 \\ 0 & 1 \end{vmatrix} u(t),$$

$$y_r(t) = \begin{vmatrix} 0.0021 & -0.0031 \\ -0.0026 & 0.0037 \end{vmatrix} z_r(t),$$

(3.14)

Where $z_r = \begin{bmatrix} z_{r1} \\ z_{r2} \end{bmatrix}$ are state variable, $u = \begin{bmatrix} dl \\ dv \end{bmatrix}$ are two manipulated inputs, and $y_r = \begin{bmatrix} dx_B \\ dx_D \end{bmatrix}$ are two outputs of LPG and gasoline product.

Stability test. The system is asymptotically stable since all eigenvalues of the state matrix are in the left half of the complex plane (([−0.5105, −0.5105]).

MRAC BUILDING AND SIMULATION

Adaptive control system is the ability of a controller which can adjust its parameters in such a way as to compensate for the variations in the characteristics of the process. Adaptive control is widely applied in petroleum industries because of the two main reasons: firstly, most of processes are nonlinear and the linearized models are used to design the controllers, so that the controller must change and adapt to the model-plant mismatch; secondly, most of the processes are nonstationary or their characteristics are changed with time, and this leads again to adapt the changing control parameters.

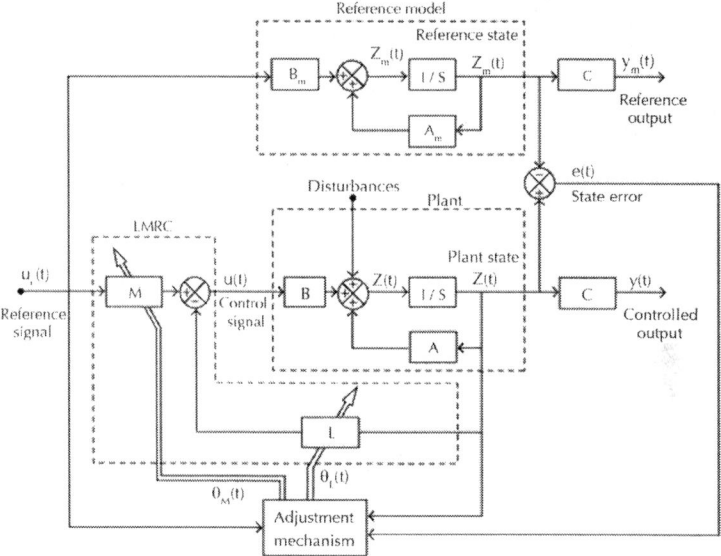

Figure 4: MRAC block diagram.

The general form of an MRAC is based on an inner-loop Linear Model Reference

Controller (LMRC) and an outer adaptive loop shown in Figure 4. In order to eliminate errors between the model and the plant and the controller is asymptotically stable, MRAC will calculate online the adjustment parameters in gains L and M by $\theta_L(t)$ and $\theta_M(t)$ as detected state error $e(t)$ when changing A, B in the process plant.

Simulation program is constructed using Maltab Simulink with the following data.

Process Plant

$$\dot{z} = Az + Bu + \text{noise},$$

$$y = Cz,$$

(4.1)

Where

$$A_m = \begin{bmatrix} 0.2616 & 0 \\ 0 & 0.2616 \end{bmatrix}, B_m = \begin{bmatrix} 1 & 0 \\ 0 & 1 \end{bmatrix}, C_m = \begin{bmatrix} 0.004 & -0.007 \\ -0.0011 & 0.0017 \end{bmatrix}$$

,

$\alpha1$, $\alpha2$, $\beta1$, $\beta2$ are changing and
on the process dynamics.

Reference Model

$$\dot{z}_m = A_m z_m + B_m u_c,$$
$$y_m = C_m z_m,$$

(4.2)

where , ,

State Feedback

$$u = M u_c - L z,$$

(4.3)

where $L = \begin{bmatrix} \theta_1 & 0 \\ 0 & \theta_2 \end{bmatrix}$ and $M = \begin{bmatrix} \theta_3 & 0 \\ 0 & \theta_4 \end{bmatrix}$

Closed Loop

$$\dot{z} = (A - BL)z + BMu_c = A_c(\theta)z + B_c(\theta)u_c$$

(4.4)

Error Equation

$$\dot{e} = \dot{z} - \dot{z}_m = Az + Bu - A_m z_m - B_m u_c = A_m e + (A_c(\theta) - A_m)z + (B_c(\theta) - B_m)u_c$$
$$= A_m e + \Psi\left(\theta - \theta^0\right),$$

(4.5)

Where $\Psi = \begin{bmatrix} -\beta_1 z_1 & 0 & \beta_1 u_{c1} & 0 \\ 0 & -\beta_2 z_2 & 0 & \beta_2 u_{c2} \end{bmatrix}.$

Lyapunov Function

$$V(e,\theta) = \frac{1}{2}\left(\gamma e^T P e + \left(\theta - \theta^0\right)^T\left(\theta - \theta^0\right)\right),$$

(4.6)

where γ is an adaptive gain and P is a chosen positive matrix.

Derivative Calculation of Lyapunov Function

$$\frac{dV}{dt} = -\frac{\gamma}{2}e^T Q e + \left(\theta - \theta^0\right)^T \left(\frac{d\theta}{dt} + \gamma \Psi^T P e\right),$$

(4.7)

where $Q = -A^T_{\ m} P - PA_m$.

For the stability of the system, $dV/dt < 0$, we can assign the second item $(\theta - \theta^0)^T (d\theta/dt) + \gamma^T Pe) = 0$ or $d\theta/dt = -\gamma^T Pe$. Then we always have $dV/dt = -(\gamma/2)e^T Qe$. If we select a positive matrix $P > 0$, for instance, P

$= \begin{bmatrix} 1 & 0 \\ 0 & 2 \end{bmatrix}$, then we have $Q = -A^T_m P - PA_m = \begin{bmatrix} 0.5232 & 0 \\ 0 & 1.0465 \end{bmatrix}$ Since matrix Q is obviously positive definite, then we always have $dV/dt = -(\gamma/2)$ $e^T Qe < 0$ and the system is stable with any plant-model mismatches.

Parameters Adjustment

$$\frac{d\theta}{dt} = -\gamma \begin{bmatrix} -\beta_1 z_1 & 0 \\ 0 & -\beta_2 z_2 \\ \beta c_1 u_1 & 0 \\ 0 & \beta_2 u_{2c} \end{bmatrix} [P] \begin{bmatrix} e_1 \\ e_2 \end{bmatrix} = \begin{bmatrix} d\theta_1/dt \\ d\theta_2/dt \\ d\theta_3/dt \\ d\theta_4/dt \end{bmatrix} = \begin{bmatrix} \gamma\beta_1 z_1 e_1 \\ 2\gamma\beta_2 z_2 e_2 \\ -\gamma\beta_1 u_{c1} e_1 \\ -2\gamma\beta_2 u_{c2} e_2 \end{bmatrix}.$$

(4.8)

Simulation Results and Analysis

We assume that the reduced-order linear model in (3.14) can also maintain the similar steadystateoutputs as the basic nonlinear model. Now we use this model as an MRAC to take theprocess plant from these steady-state outputs ($x_D = 0.9654$ and $x_B = 0.0375$) to the desiredtargets ($0.98 \leq x_D \leq 1$ and $0 \leq x_B \leq 0.02$) amid the disturbances and the plant-modelmismatches as the influence of the feed stock disturbances. The design of a new adaptive controller is shown in Figure 5 where we install an MRAC and a closed-loop PID (Proportional, Integral, Derivative) controller to eliminate theerrors between the reference setpoints and the outputs.

We run this controller system with different plant-model mismatches, for instance, aplant with $A = \begin{bmatrix} -0.50 & 0 \\ 0 & -0.75 \end{bmatrix}$, $\begin{bmatrix} 1.5 & 0 \\ 0 & 2.5 \end{bmatrix}$ and an adaptive gain $\gamma = 25$. The operating setpointsfor the real outputs are $x_{DR} = 0.99$ and $x_{BR} = 0.01$. Then, the reference setpoints for the PIDcontroller are $r_D = 0.0261$ and $r_B = -0.0275$ since the real steady-state outputs are $x_D = 0.9654$ and $x_B = 0.0375$. Simulation in Figure 6 shows that the controlled outputs x_D and x_B area lways stable and tracking to the model outputs and the reference setpoints (the dotted lines,r_D and r_B) amid the disturbances and the plant-model mismatches.

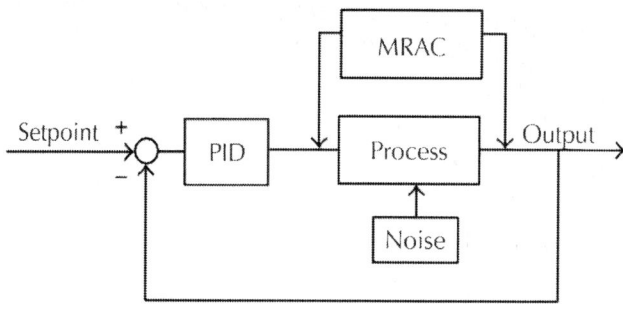

Figure 5: Adaptive controller with MRAC and PID.

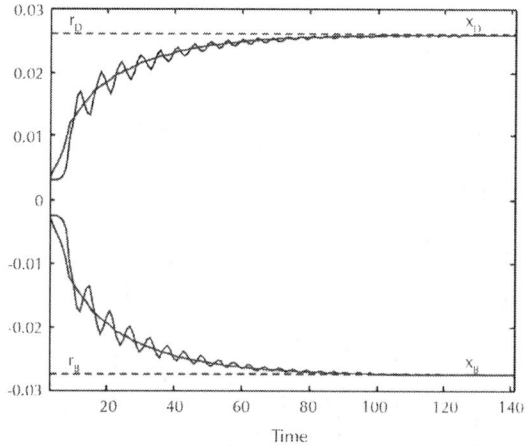

Figure 6: Correlation of plant outputs, model outputs, and reference setpoints.

CONCLUSIONS

We have introduced a procedure to build up a mathematical model and simulation for a condensate distillation column based on the energy balance (L-V) structure. The mathematical modeling simulation is accomplished over three phases: the basic nonlinear model, the full-order linearized model and the reduced-order linear model. Results from the simulations and analysis are helpful for initial steps of a petroleum project feasibility study and design.

The reduced-order linear model is used as the reference model for an MRAC controller. The controller of MRAC and PID theoretically allows the plant outputs tracking the reference setpoints to achieve the desired product quality amid the disturbances and the model-plant mismatches as the influence of the feed stock disturbances.

In this paper, the calculation of the mathematical model building and the reduced-order linear adaptive controller is only based on the physical laws from the process. The real system identifications including the experimental production factors, specific designed structures, parameters estimation, and the system validation are not mentioned here. Further, the MRAC controller is not suitable for the on-line handling of the process constraints.

REFERENCES

1. PetroVietnam Gas Company, "Condensate processing plant project—process description," Tech. Rep. 82036-02BM-01, PetroVietnam, Washington, DC, USA, 1999.

2. E. Marie, S. Strand, and S. Skogestad, "Coordinator MPC for maximizing plant throughput," Computers & Chemical Engineering, vol. 32, no. 1-2, pp. 195–204, 2008. ·

3. H. Kehlen and M. Ratzsch, "Complex multicomponent distillation calculations by continuous thermodynamics," Chemical Engineering Science, vol. 42, no. 2, pp. 221–232, 1987. ·

4. R. G. E. Franks, Modeling and Simulation in Chemical Engineering, Wiley-Interscience, New York, NY, USA, 1972.

5. W. L. Nelson, Petroleum Refinery Engineering, McGraw-Hill, Auckland, New Zealand, 1982.

6. M. V. Joshi, Process Equipment Design, Macmillan Company of India, New Delhi, India, 1979.

7. W. L. McCabe and J. C. Smith, Unit Operations of Chemical Engineering, McGraw-Hill, New York, NY, USA, 1976.

8. P. Wuithier, Le Petrole Raffinage et Genie Chimique, Paris Publications de l'Institut Francaise du Petrole, Paris, France, 1972.

9. G. Stephanopoulos, Chemical Process Control, Prentice-Hall, Englewood Cliffs, NJ, USA, 1984.

10. O. Katsuhiko, Model Control Engineering, Prentice-Hall, Englewood Cliffs, NJ, USA, 1982.

11. A. Papadouratis, M. Doherty, and J. Douglas, "Approximate dynamic models for chemical process systems," Industrial & Engineering Chemistry Research, vol. 28, no. 5, pp. 522–546, 1989.

12. S. Skogestad and M. Morari, "The dominant time constant for distillation columns," Computers & Chemical Engineering, vol. 11, no. 7, pp. 607–617, 1987. ·

Quality of Electroless Ni-P (Nickel-Phosphorus) Coatings Applied in Oil Production Equipment with Salinity

Fernando B. Mainier, Maria P. Cindra Fonseca,
Sérgio S. M. Tavares, and Juan M. Pardal

Escola de Engenharia, Universidade Federal Fluminense (UFF), Niterói,
Brazil

ABSTRACT

The corrosion resistance of nickel-phosphorus (Ni-P) coatings and their mechanical properties in seawater have led investigations into the development of new technologies and the replacement of some special alloys in equipment used in oil production, such as valves, tubing, sucker rod joints, pumps, riser, manifolds and subsea Christmas trees. These studies began with Brenner and Riddel who developed, in the 1940s, formulations for Ni-P deposition on carbon steel without using an electric current. Joint deposition of nickel and phosphorus on a metallic surface (carbon steel) without applying an external current is accomplished using cathodic reduction with hydrogen (H) from a reducing agent (sodium hypophosphite) and nickel salts. To assure good performance of a Ni-P coating, the deposit quality must be inspected and evaluated during the chemical deposition process or in the end product. The recommended test parameters are: thickness, layer uniformity, hardness, adhesion, porosity, corrosion resistance and chemical composition of the nickel-phosphorus coating. The purpose of this paper was to investigate the Ni-P coating process, to evaluate the behaviour of Ni-P in a saline environment using aqueous brine (3.5% - 30% sodium chloride by mass) and to present possible defects that could compromise the coating.

INTRODUCTION

One of the traditional techniques to maintain the mechanical characteristics of a material in the manufacturing of industrial equipment, primarily using carbon steel or low alloy steels, as well as to make surfaces more resistant to abrasion and corrosion, is without doubt, by applying a specific finish, such electroless nickel-phosphorus plating (Ni-P). The Ni-P coating is deposited on carbon steel without the application of an external electrical current. This feature has therefore led to, directly or indirectly, the development of special tools and new technologies in the area of oil production in a high salinity environment associated with corrosive gases such as CO_2 and H_2S.

Historically, the process of nickel deposition was initiated in 1844 by the work of Wurtz [1,2], who discovered the reduction of Ni^{2+} ions to metallic porous nickel (Ni), and subsequently by Brenner and Riddel [3], who over one hundred years later developed formulations and practices for Ni-P deposition on carbon steel without the aid of an electric current. Studies by Duncan [4], Colaruotolo [5], Mainier et al. [6], Tallinn [7], Weil et al. [8], Mainier and Araújo [9], Delaunois et al. [10], Liu et al. [11] and Baudrand [12] have shown that the rate of growth and application of Ni-P coatings since the 1980s has increased in several industrial sectors.

The performance of this finish has led to its use in various industrial areas such as the production of pulp and paper, plastics, petrochemicals, textiles, automobiles, aeronautics, electronics and food. Also, faced with high corrosivity and the increasing challenges of obtaining petroleum under adverse conditions have driven the oil industry to seek new alternatives in the areas of materials. Ni-P coatings, due to their abrasion and corrosion resistance, have presented excellent performance, mainly in terms of valves, special tools, risers, pumps and production pipes. Figure 1 shows some of these applications in the petroleum industry.

The technical literature and patents concerning nickel deposition processes indicate a wide range of development in terms of new formulations, additives and other co-precipitations with cobalt (Co), boron (B), silicon carbide (SiC), Teflon, etc., which will provide new properties to Ni-P coatings and therefore lead to new Industrial applications.

DEPOSITION PROCESS OF NICKEL-PHOSPHORUS

The electrochemical deposition process, without the aid of an electric current, is presented in Figures 2 and 3. The process starts with the receipt of parts in quality control where they are formalized according to operational procedures that should assess the following parameters: dimensional analysis, definition and assessment of the constituent materials of the part, rules and procedures of surface preparation (chemical or mechanical cleaning for removal of oxides, oily material or grease), the type of bath to be used, temperature, speed of deposition, thickness, and the nickel/phosphorus ratio, among others.

MECHANISM AND PROPERTIES OF NI-P DEPOSITION

In the conventional process of nickel electroplating, Ni^{2+} ions (present in the bath) are reduced to metallic nickel by an external electric current and deposited onto the surface of a metallic material connected to the negative pole (cathode), while the positive pole (anode) is usually made up of high purity nickel, as shown in the illustration in Figure 4. The thickness of the deposited nickel film and its properties depend on the electric current density, the voltage in volts applied, the concentration of salts, the bath temperature, pH, the nature of the base metal (cathode) and the additives used to give specific features to the nickel coating.

Figure 1: Ni-P coating process in pipeline valve, production pipe and the interior of a corrosive gas storage cylinder.

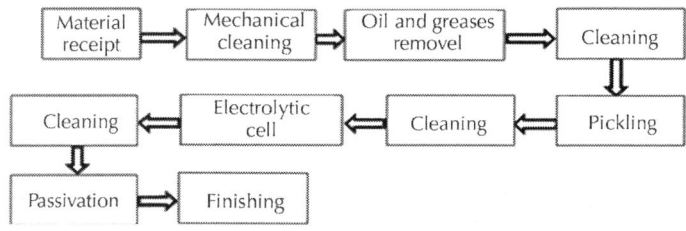

Figure 2: Deposition process.

Figure 3: Overview of the bath of nickel-phosphorus.

The first point that differentiates this process is that the deposition of autocatalytic chemical nickel plating requires no external electric current; that is, the process is self-regulated by the kinetics of the reactions involved. Continuous, uniform joint co-deposition of nickel and phosphorus is achieved by cathodic reduction with atomic hydrogen (H) produced in the bath from the hydrolysis of the reducing agent (NaH_2PO_2). The baths used in the autocatalytic process are more complex and require more control, generally, and are formulated based on nickel salts (Ni^{2+}), the reducing agent (NaH_2PO_2) and additives that control the pH, complexing and the addition of other salts to ensure the quality of the coating [13].

The required scientific and technological knowledge, as well as the reaction mechanisms to explain the deposition of Ni-P, are credited to Gould et al. [14], Duncan [4], Belinsky [15], Mallory [16] and Riedel [17], among others. It can be assumed, on the basis of these authors, that the coating deposition kinetics of Ni-P is based on the following main points and shown in Figure 5.

- Atomic hydrogen capacity formation;
- Hydrogen adsorption capacity by the metal surface;
- Reduction of Ni^{2+} ions and sodium hypophosphite;
- Co-deposition of nickel and phosphorus on the metal surface.

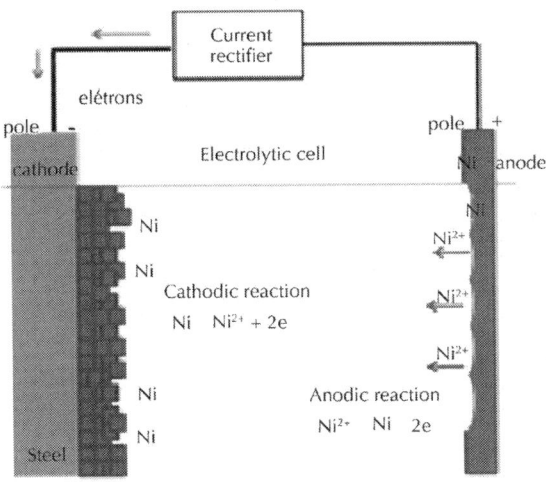

Figure 4: Illustration of electrolytic deposition of nickel on carbon steel.

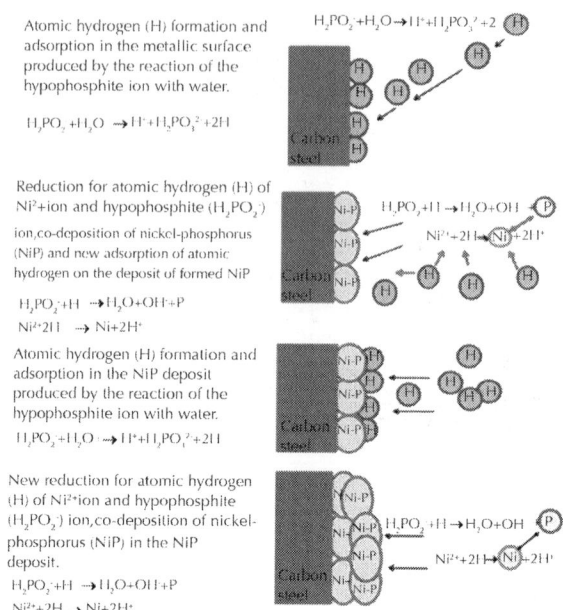

Atomic hydrogen (H) formation and adsorption in the metallic surface produced by the reaction of the hypophosphite ion with water.

$H_2PO_2 + H_2O \rightarrow H^+ + H_2PO_3^- + 2H$

Reduction for atomic hydrogen (H) of Ni^{2+} ion and hypophosphite ($H_2PO_2^-$)

ion, co-deposition of nickel-phosphorus (NiP) and new adsorption of atomic hydrogen on the deposit of formed NiP

$H_2PO_2^- + H \rightarrow H_2O + OH^- + P$
$Ni^{2+} + 2H \rightarrow Ni + 2H^+$

Atomic hydrogen (H) formation and adsorption in the NiP deposit produced by the reaction of the hypophosphite ion with water.
$H_2PO_2 + H_2O \rightarrow H^+ + H_2PO_3^- + 2H$

New reduction for atomic hydrogen (H) of Ni^{2+} ion and hypophosphite (H_2PO_2) ion, co-deposition of nickel-phosphorus (NiP) in the NiP deposit.
$H_2PO_2^- + H \rightarrow H_2O + OH^- + P$
$Ni^{2+} + 2H \rightarrow Ni + 2H^+$

Figure 5: Mechanism of Ni-P layer formation of on carbon steel.

In the view of Etcheverry [18], the co-deposition of Ni-P has the behavior of an alloy, either crystalline or amorphous, depending on the percentage (mass %) of nickel and phosphorus. Figure 6(a), below, shows the microscopic aspects of Ni-P deposition observed by scanning electron microscopy; Figure 6(b) shows an image taken with an optical microscope.

It is important to note that parallel or unexpected reactions can occur in the bath by reducing the ability of codeposition NI-P coating, such as:

- $H_2PO_2^- + H_2O \rightarrow H^+ + H_2PO_3^{2-} + H_2$—the hydrolysis of hypophosphite to form molecular hydrogen (H_2) instead of atomic hydrogen (H), which entails a decrease in the reducing power;

- $2\,H \rightarrow H_2$—the natural loss of atomic hydrogen (H) reduction capacity;

- $Ni^{2+} + H_2PO_3^{2-} \rightarrow NiHPO_3$—the possibility of nickel ion (Ni^{2+}) precipitation in the form of nickel (II) hydrogen phosphite ($NiHPO_3$), resulting in the impoverishment of the Ni^{2+} ion concentration in the bath or, if deposited in the Ni-P coating, it can make a rougher coat.

To avoid this kind of problem, it is essential that substances are present in the bath that act on complexants of Ni^{2+} ions (the preventing precipitation). For this, we used organic acid-based formulations, such as tartaric acid, malic acid, succinic acid, adipic acid, hydroxypropionic acid, etc. In addition, the concentration of hydrogen phosphite ($H_2PO_3^{2-}$) ions increases as atomic hydrogen (H) is produced. This concentration should be limited and the occurrence of hydrogen phosphite ion co-deposition along with Ni-P should be monitored as this can lead to a porous coating. Laboratory experiments have shown that the addition of activating substances such as adipic acid, succinic acid, etc. can slow the deposition speed of Ni-P and to avoid this kind of problem [17].

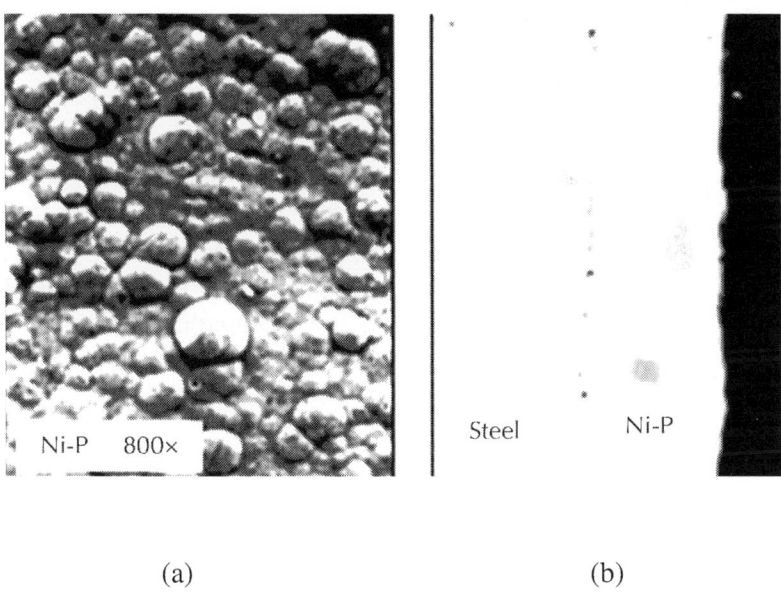

(a)   (b)

Figure 6: (a) Scanning electron microscopy (800×); (b) Optical microscopy (200×).

In this way, the deposition process is complex and, therefore, the speed of Ni-P deposition on the surface of a part, according to Riedel [17], is preferably a function of the following parameters: temperature, pH, concentration (nickel hypophosphite salts, complexants, activators, stabilizers and contaminants), surface roughness, agitation of the bath and the area/volume relationship of the electroless nickel bath.

According to Riedel [17], the temperature and the pH are important parameters since, as the deposition temperature and pH increase, in relation to most baths, an increased speed of deposition is favored, which can create voids or porosity in the deposited layer. The graphs presented in Figure 7 are based on baths containing 15 g/L nickel chloride and 10 g/L sodium hypophosphite to show this trend.

NI-P COATING: SPECIFICATIONS AND PROPERTIES

The following are some properties of Ni-P coatings

[17,19] based on the nickel and phosphorus levels (Table 1).

For a Ni-P coating applied on carbon steel come meet the adverse conditions of the production of petroleum products, it is essential to adopt a methodology critical inspection "in situ", with due allowance for the manufacturing process of the equipment and the process of deposition based on ISO 4527 [20].

In the critical inspection of process of Ni-P deposition is important to know the various routines comprising the process itself, and must be inspected, among others, the areas of preparation (sandblasting and cleaning chemistry), the process control laboratory (qualified personnel, equipment and assessment procedures), electrochemical baths and the finishing area.

In the actual inspection of the parts should be required to evaluate the following parameters: appearance, uniformity, layer thickness, abrasion, chemical composition, adhesion, porosity and hardness. In addition, with a view to the corrosivity of petroleum, containing CO_2 and H_2S [21], corrosion tests in order to increase the security of the application and the use of Ni-P coating under adverse conditions.

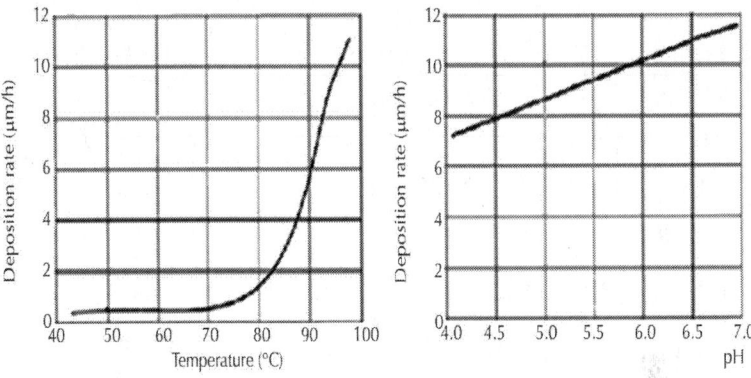

Figure 7: Speed of deposition (µm/h) of Ni-P as a function of pH and temperature.

Table 1: Properties of nickel-phosphorus coatings

Properties	Posphorus content(%)		
	Low	**Medium**	**High**
Nickel, % (mass)	96 - 99	92-95	88-91
Phosphorus, % (mass)	1-4	5-8	9-12
Vickers microhardness without heat treatment, HV	650-750	500-550	450-500
Vickers microhardness with heat treatment, HV	100-1050	900-950	850-900
Melting point, °C	1200	890	870
Density, g/cm³	8.5-8.7	8.1-8.3	7.7-7.8
Resistivity, µΩ/ cm	50	70	90

Resistance to abrasion	Superior	Very good	Very good
Weldability	Good	Regular	Bad

Appearance

On the coated parts visual inspection should not show any defects such as pitting, exfoliations, bubbles, cracks, deposits or failures that may constitute an impediment on the performance of the material.

Deposit Thickness and Uniformity

To work in high aggressive environments are recommended thicknesses ranging from 75 to 125 μm. The procedure for the determination of the thickness of Ni-P deposit must be specified by the user and/or established by common agreement between the parties, and may be used microscopic methods, magnetic, ultrasonic, etc.

Optical microscopy determines the thickness and uniformity of deposition, however, this method is destructive. To solve this type of problem is allowed the use of a test coupon (testimony) representative for measurement of coating thickness. For a comparison, the micrographs of Ni-P coating applied on carbon steel show in Figure 8, the following are examples of uniformity of layer (Figure 8(a)), while in Figure 8(b) are presented defects occurring during processing of deposition.

Chemical Composition

The corrosion resistance of Ni-P coating depends on the concentration of nickel and phosphorus present in the deposited layer and increasing the phosphorus content improves the anti-corrosion protection. The ISO 4527 [20] standard presents in the Table 2 the chemical composition acceptable for Ni-P coatings. The determination of nickel and phosphorus content in the Ni-P deposit can be made using the following analytical techniques: wet chemical, atomic absorption, x-ray fluorescence, plasma, etc. For the severe conditions of industrial use the phosphorous content shall not be less than 10% by mass.

Porosity

The porosity of the electroless Ni-P coating is a most important parameter. The deposition of nickel-phosphorus should be free of porosity in order to prevent the corrosive medium contact with the base metal (carbon steel), the corrosive process begins, often through pores or failures. This problem becomes dangerous when the metal is anodic in the relation to Ni-P deposit, resulting in a galvanic cell (galvanic corrosion).

In the evaluation of Ni-P coating porosity is used the Ferroxyl Method [20,22]. This method is to place a filter paper on the piece and then applies a mixture of solution $K_3Fe(CN)_6$—potassium ferricyanide and sodium chloride on the piece during 30 seconds.

The appearance of blue points indicates the porosity and the attack on the base metal (carbon steel) as shown, then, Figure 9. The reactions: $Fe-2e \rightarrow Fe^{2+}$ correspond, to attack the carbon steel and the blue color development on the filter paper placed on the piece or test coupon.

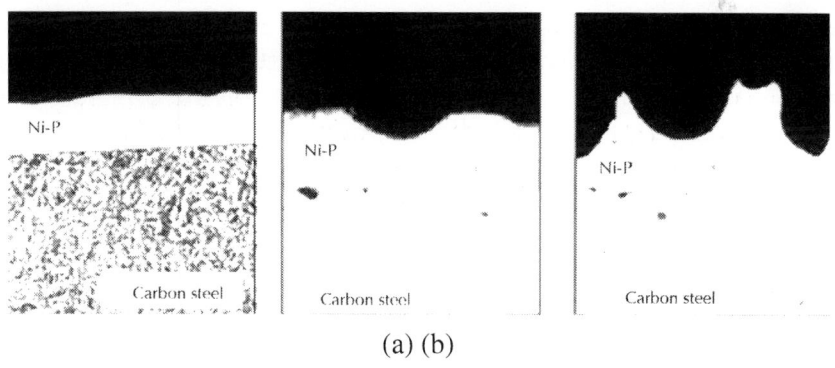

(a) (b)

Figure 8: Optic micrographs of Ni-P coating applied on carbon steel: (a) Uniform; (b) Defective layers.

Table 2: Chemical composition of Ni-P deposit [20]

Elements	Chemical Composition(%mass)		
	Minimum	Maximum	Typical

Nickel	85	98	88-95
Phosphorus	2	15	5-12
Other (Al, As, B, Bi, C, Cd, Co, Cr, Cu, Fe, Mn, Nb, Pb, S, Sb, Se ,Si, Sn, V, Zn)	0	2.0	0.05

$$3Fe^{2+} + 2\left[Fe(CN)_6\right]^{3-}$$
$$\rightarrow Fe_3\left[Fe(CN)_6\right]_2 (\text{blue points})$$

Microhardness

The hardness of Ni-P deposit indicates whether part or not specific heat treatment suffered and usually ranges from 500 to 580 HV (Vickers microhardness). After heat treatment depending on the exposure time and temperature the hardness can vary from 600 to 1100 HV. The heat treatment applied to Ni-P coatings shall comply with the directions of the rules in order to minimize the occurrence of cracks or fissures [20,23].

Agarwala & Agarwala [24] show that the types of baths determine the relationships of the nickel and phosphorus levels in the deposit and the increase of the levels of phosphorus reduces the hardness with and without heat treatment as shown in the Table 3.

Adhesion

The adhesion of electroless Ni-P to non-conductors is dependent on mechanical keying with associated Van des Waals force. Ni P coatings have good adherence to carbon steel due to the fact that the cohesion of the forces on the metal-base film are often in superior of 140 MPa (20,000 psi). The evaluation of this adhesion is essential and often so that the tests are acceptable, it is indispensable to carry out various tests. These tests are usually comparative, qualitative and quantitative tests are based on an aluminum disk collage on the metallic surface and applying a continuous tension force in aluminum disk until the breakup. However, its implementation depends on the geometry of the piece and its use. Adhesion tests are based on standards: ASTM B 733 [25] and ISO 4527 [20].

Abrasion

Resistance to abrasion is directly related to the phosphorous content, heat treatment and adherence to the surface of the base metal. In general, the increase of the levels of phosphorus and the increase of hardness provided by heat treatment increases the abrasion resistance. Abrasion tests are specified and depending on the specific use of electroless Ni-P coating as recommends ISO 4527 [20].

Corrosion

In evaluating the performance of Ni-P coating in different aggressive media are used in laboratory testing and field. In the case of production of oil this Ni-P coating has proven quite attractive, with a view to its good resistance to many corrosive media such as corrosive gases (CO_2, H_2S) and high water salinity, contaminants commonly found in petroleum. However, it is essential to establish two factors are important in corrosion resistance: the phosphorous content which should be more than 10 % (mass) and the thickness shall not be less than 75 mm.

Electroless Ni-P coating are not recommended for some corrosive media containing chloride based compounds and ferric sulfate, nitrates, nitrites and ammonium compounds. Laboratory testing to evaluate the performance of electroless Ni-P coating in petroleum with high aggressiveness can be static or dynamic.

The high aggressiveness can be represented by mixes with salt water and pressurized using mixtures of CO_2, H_2S. The corrosion test can be carried out in a pressurized cell to 70 kgf/cm^2. The temperature can range from 25°C to 80°C and exposure time of 240 hours continuous hours.

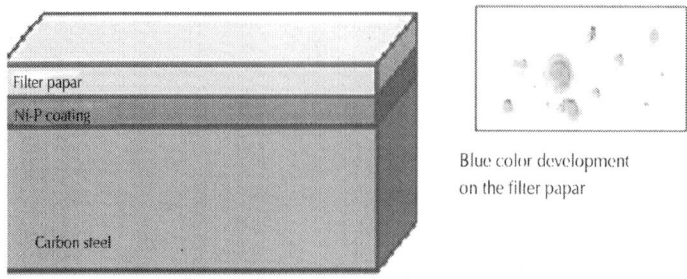

Blue color development
on the filter papar

Figure 9: Ferroxyl test applied to determine the porosity of Ni-P deposit on carbon steel.

Table 3: Deposit hardness of Ni-P-carbon steel [24]

Phosphorus,% (mass)	Vickers microhardness(HV)	Vickers microhardness with heat treatment, HV
2-3	700	1000
6-9	550	920
10-12	510	880

CORROSION TESTING OF ELECTROLESS NI-P COATING

The carbon steel coupons coated Ni-P, heavy, previously were placed in pressurized containers (70 kgf/cm²) with 300 mL containing sodium chloride solutions. Sodium chloride solutions used in the test were, respectively, of 3.5%, 10%, 20% and 30% by weight. The tests were run temperature of 25°C and for 250 hours of total immersion. The properties of coupons with Ni-P coating are presented below in Table 4.

After the end of the coupons were washed with running water, alcohol and dried with hot air and some coupons were cut to evaluate by optical metallography the corrosive effects of the salt solution.

The tests performed with Ni-P coated coupons did not show noticeable losses in mass and the analysis by optical microscopy did not find attacks on coupons, i.e. it can be considered that the corrosion rates are void.

CONCERNS OR PRECAUTIONS REGARDING NICKEL-PHOSPHORUS COATING APPLIED ON CARBON STEEL AND USED IN EXTREME CONDITIONS

Generating ideas of the Principles of Precaution, probably born in the 1970 with the fledgling company's concerns with the ethics, with the environment, with the deterioration of water resources, with the risk of contamination, with the uncertainties of processed foods and the insecurities of the technological applications in various industrial segments.

The term precaution, in the vision of sustainability and of triad industrial safety-environment-workers' health and their descendants, should mean taking measures to protect human health and the environment from possible damage that may happen. This means that the processes must be safe to avoid or minimize possible disasters, however, in the international area; there are different views that caution has different interpretations, especially when the more industrialized countries impose products and/or obsolete technologies to less developed countries.

Table 4: Properties of Ni-P coated coupons

Properties	Evaluation
Phosphorus, %(mass)	10.5
Thickness(μm)	75
Porosity	Free
Apparent defects	Free

Adhesion	Very good
Vickers microhardness(HV)	600

In the view of Mainier [26] systems productive, knowledgeable of the risks of industrial manufacturing processes and seeming not to care about the present and not the future, continue to exert strong pressure on the environment, imposing or masking obsolete technologies that include waste, packaging, recycling and toxic waste, themes that often get confused or are linked.

On the interests and economic philosophies and large industrial manufacturing complexes and industrialized countries become fellow-agents of a policy of mutual interest and, in many situations, against the interests of the man himself. In this optic, spills, leaks and contamination with large environmental impact have taken place.

Under the technical point of view can be worrying speed of deposition of Ni-P coating for high thickness in relation to the final cost of the coated part. For example, hypothetically, assuming a thickness 75 μm and with reference to three electrochemical baths, respectively, with 3 μm/h, 10 μm/h and 25 μm/h minimum operating time for these operations would be, respectively, 25 h, 7.5 h and 3 h.

In a simplistic evaluation the first case will probably have a longer time, forcing the continuous control and systematic and will have higher quality of landfill although the cost is higher when compared to other procedures. At a high speed the possibility of porosity and faults can occur with greater intensity when compared with low layer deposition speed of Ni-P.

Another point that also needs to be evaluated is the life cycle of the piece and its implications in the project as a whole when compared to application of a coating with other leagues more massive and noble and give greater security to the investment.

CONCLUSIONS

On the basis of the facts stated it can be concluded that:

- To ensure the good performance of electroless Ni-P coating is necessary to know the process of electrochemical deposition without external current and secure inspection philosophies during processing and in the final product.

- The quality of Ni-P coating must be supported for constant qualification standards and national and international procedure.
- It is essential to analyze the speed of deposition of coatings on the basis of cost and the possibility of defects to be used in extreme conditions.
- Laboratory testing showed excellent performance for Ni-P coatings with 75 μm in salt solutions of sodium chloride 3.5% to 30% by mass.

REFERENCES

1. A. C. R. Wurtz, "On Copper Hydride," Hebdomadaires des Séances de l'Académie des Sciences, Vol. 18, 1844, pp. 702-704.

2. A. C. R. Wurtz, "On Copper Hydride," Hebdomadaires des Séances de l'Académie des Sciences, Vol. 21, 1845, p. 149.

3. A. Brenner and G. E. Riddel, "Nickel Plating on Steel by Chemical Reduction," Journal of Research of the National Bureau of Standards, Vol. 37, No. 1, 1946, pp. 31- 34.http://dx.doi.org/10.6028/jres.037.019

4. R. N. Duncan, "Performance of Electroless Nickel Coated Steel in Oil Field Environments," Material Performance, Vol. 21, 1983, pp. 28-34.

5. J. F. Colaruoto, B. V. Tilak and R. S. Jasinki, "Corrosion Charactheristcs of Electroless Nickel Coating of Oil Field Environments," Proceedings of Electroless Nickel Conference IV, Chicago, 22-24 April 1985.

6. F. B. Mainier, I. M. R. A. Brüning and E. F. Pamplona, "Desenvolvimento de Recipientes para Acondicionamento de Gás Natural Contendo Gases Corrosivos," In: VI Encontro Brasileiro de Tratamento de Superfícies (EBRAT-1989), ABTS, São Paulo, 1987, pp. 66-81.

7. V. T. Talinn, "In the World of Electroless Nickel," Finishing, Vol. 12, 1988, p. 26.

8. R. Weil, J. H. Lee and K. Parker, "Comparison of Some Mechanical and Corrosion Properties of Electroless and Electroplated Nickel

Phosphorus Alloys," Plating and Surface Finishing, Vol. 76, 1989, pp. 62-66.

9. F. B. Mainier and M. M. Araújo, "On the Effect of the Electroless Nickel-Phosphorus (Ni-P) Coating Defects on the Performance of This Type of Coating in Oilfield Environments," SPE Advanced Technology Series, Vol. 2, No. 1, 1994, pp. 63-67.

10. F. Delaunois, J. P. Petitjean, P. Lienard and M. JacobDuliere, "Autocatalytic Electroless Nickel-Boron Plating on Light Alloys," Surface and Coatings Technology, Vol. 124, No. 2-3, 2000, pp. 201-209.

11. X. Liu, J.-Q. Gao and W.-B. Hu, "Application of Electroless Ni-P Alloys in Electronic Industry," Plating & Finishing, Vol. 28, No. 1, 2006, pp. 30-34.

12. D. Baudrand, "Adhesion of Electroless Nickel Deposits to Aluminum Alloys—We Now Have a Better Understanding of the Factors Influencing Adhesion," Products Finishing, Vol. 63, No. 10, 2009, pp. 80-87.

13. ASTM B-656, "Auto Catalytic (Electroless) Nickel-Phosphorus Deposition on Metals for Engineering Use," American Society for Testing and Materials, West Conshohocken, 1992.

14. A. Gould, P. J. Boden and S. J. Harris, "Phosphorus Distribution in Electroless Nickel Deposits," Surface Technology, Vol. 12, No. 1, 1981, pp. 93-102.http://dx.doi.org/10.1016/0376-4583(81)90140-0

15. J. Bielinski, "The Role of Buffers and Complex Formers in Electroless Nickel Plating," Oberflache Surface, Vol. 25, No. 12, 1984, pp. 423-429.

16. G. O. Mallory and J. B. Hadju, "Electroless Plating: Fundamentals & Applications," Cambridge University Press, Cambridge, 1990.

17. W. Riedel, "Electroless Nickel Plating," Redwood Press Limited, Liverpool, 1991.

18. B. Etcheverry, "Adhérence, Mécanique et Tribologie des Revêtements Composites NiP—Talc Multifonctionnels à Empreinte Écologique Réduite," Institut National Polytechnique de Toulouse, Toulouse, 2006.

19. R. P. Tracy and G. J. Shawham, "Pratical Guide to Using Ni-P Electroless Nickel Coatings," Materials Performance, Vol. 29, No. 7, 1990, pp. 65-70.

20. ISO-4527, "Autocatalytic Nickel-Phosphorus Coatings, Specification and Test Method," International Organization for Standardization, Geneva, 1987.

21. F. B. Mainier, G. C. Sandres and R. J. Mainier, "Integrated Management System for In-House Control of Accidental Hydrogen Sulfide Leaks in Oil Refineries," International Journal of Science and Advanced Technology, Vol. 2, No. 9, 2012, pp. 76-84.

22. ASTM B-656, "Auto Catalytic (Electroless) Nickel-Phosphorus Deposition on Metals for Engineering Use", American Society for Testing and Materials, West Conshohocken, 1992.

23. ASTM B-578, "Measurements of Micro Hardness of Electroplated Coating," American Society for Testing and Materials, West Conshohocken, 1992.

24. R. C. Agarwala and V. Agarwala, "Electroless Alloy/ Composite Coatings: A Review," Sadhana, Vol. 28, No. 3-4, 2003, pp. 475-493.

25. ASTM B-733, "Standards Specifications for Autocatalytic Nickel-Phosphorus Coatings on Metals," American Society for Testing and Materials, West Conshohocken, 1992.

26. F. B. Mainier, "Uma Visão Crítica das Rotas Industriais de Fabricação de Produtos Químicos Utilizados nos Tratamentos de Água," Congresso de Equipamento e Automação da Indústria Química, Associação Brasileira da Indústria Química (ABQUIM), São Paulo, 1999.

Chapter 6

Pre-Treatment of High Free Fatty Acids Oils by Chemical Re-Esterification for Biodiesel Production—A Review

Godlisten G. Kombe[1], Abraham K. Temu[1], Hassan M. Rajabu[2], Godwill D. Mrema[1], Jibrail Kansedo[3], and Keat Teong Lee[3]

[1]Department of Chemical and Mining Engineering, College of Engineering and Technology, University of Dar es Salaam, Dar es Salaam, Tanzania

[2]Department of Energy Engineering, College of Engineering and Technology, University of Dar es Salaam, Dar es Salaam, Tanzania

[3]School of Chemical Engineering, Universiti Sains Malaysia, Pinang, Malaysia

ABSTRACT

Non edible oil sources have the potential to lower the cost of biodiesel. However, they usually contain significant high amounts of free fatty acids (FFA) that make them inadequate for direct base catalyzed transesterification reaction (where the FFA content should be lower than 3%). The present work reviews chemical re-esterification as a possible method for the pre-treatment of high FFA feedstock for biodiesel production. The effects of temperature, amount of glycerol, type and amount of catalyst have been discussed. Chemical re-esterification lowers FFA to acceptable levels for transesterification at the same time utilizing the glycerol by product from the same process. Further researches have been proposed as a way forward to improve the process kinetics and optimization so as to make it more economical.

INTRODUCTION

The worldwide worry about the protection of environment and the dependence on fossil fuel has given rise to development of alternative energy sources as substitute for traditional fossil fuels. Fossil fuel sources are nonrenewable, and will be exhausted in the near future. According to Alekett [1] the world's oil reserves are up to 80% less than predicted, which calls for alternative sources of energy. Biodiesel is one of the renewable energy fuel sources alternatives to the conventional petroleum diesel. It is simply produced by transesterification process whereby the vegetable oil or animal fat (Triglyceride) reacts with alcohol in presence of catalyst or without catalyst to give the corresponding alkyl esters of the fatty acid mixture that is found in the parent vegetable oil or animal fat [2,3]. Transesterification reaction can be uncatalyzed, base-catalyzed, acid-catalyzed or enzymecatalyzed.

Research on biodiesel production has captured the attention of different researchers with focus on heterogeneous catalysts which have received considerable attention [4-9]. Unfortunately these studies did not lead to the development of catalysts with high activity, good reusability and stability in order to replace the homogenous catalysts, such as the sodium methoxide or hydroxide [10-12]. Today, homogeneous base-catalyzed transesterification process is widely used industrially due to the fact that, it is kinetically much faster and it has been proven to be economically viable [13]. According to the report by Bacovsky, Körbitz, Mittelbach and Wörgetter [14] on the status of biodiesel production technology, most of the commercialized biodiesel production technology utilizes homogeneous base-catalyzed transesterification. However, the main drawback of this technology is its sensitivity to the purity of feedstocks, especially water and free fatty acid content [15-17].

The use of edible grade oils as feedstocks competes with food supply in the long-term [18] and accounts for the higher price of biodiesel, since the cost of raw materials accounts for 60% to 80% of the total cost of biodiesel fuel [19-22]. One way of reducing the biodiesel production costs is to use the less expensive feedstocks mostly containing fatty acids such as inedible oils, animal fats, waste food oil and by-products of the refining vegetable oils [23-25]. However, feedstocks high in free fatty acid are not easily converted by homogeneous base transesterification, because of the concurrent soap formation of the free fatty acids with the catalyst. The excessive amount of soap formed significantly interferes with the washing process by forming emulsions, thus leading to substantial yield losses [26-30]. The reaction can only tolerate FFA content up to 3% in the feedstock without affecting the process negatively as suggested by Knothe, Van Gerpen and Krahl [31]. The free fatty acid (FFA) value lower than 3% is recommended for higher conversion efficiency [32]. The pre-treatments of non-edible oils for lowering the FFA in feedstock for base catalyzed transesterification are therefore inevitable.

The pre-treatment of high FFA with acid catalysis followed by base catalyzed transesterification has been proposed by several authors [2,23,33,34]. The process can lower a high FFA feedstock to ≤0.5% quickly and effectively. However, depending on the amount of FFA in the oils or fats; one-step pre-treatment may sometimes not reduce the FFA efficiently because of the high content of water produced during the reaction [35]. In this case, a mixture of alcohol and sulphuric acid can be added into the oils or fats three times (three-step preesterification) and the water must be removed before transesterification [36]. Van Gerpen, Shanks, Pruszko, Clements and Knothe [37] suggested the use of high molar ratios of alcohol to oil as high as 40:1 to dilute the water formed during pre-treatment, yet this will require more energy to recover the excess alcohol used. The water formed during the pre-treatment phase requires removal and the use of corrosive nature or catalysts commonly H_2SO_4 which requires high capital intensive reactors, has limited the application of the process.

The chemical re-esterification (glycerolysis) process has the capability of converting the free fatty acid back to their respective glyceride molecule. This technique involves adding glycerol to the high FFA feedstock and heating it to temperature of about (200°C), with a metallic catalyst such as zinc chloride and zinc dust or without catalyst. The glycerol reacts with the FFAs to form monoglycerides, diglycerides and triglycerides [38]. It produces a low FFA feed that can be processed to methylesters using traditional homogeneous base transesterification technique. The advantage of this approach is that no alcohol is needed during the pre-treatment and the water formed from the reaction can be immediately vaporized and vented from the reaction mixture. The process has also the potential of utilizing glycerol, a by-product from transesterification and thereby lowers the cost of biodiesel. However, the drawbacks of this method are its high temperature requirement and relatively slow reaction rate [37]. Although chemical re-esterification has a potential to lower the high FFA feedstock for homogeneous base catalyzed transesterification. The re-esterification method has not been studied from the standpoint of the extent of de-acidification for biodiesel production. The literatures on the applicability of this process as a high FFA pre-treatment for biodiesel are hard to find. It is the purpose of this paper to review the applicability of the glycerolysis process as a pre-treatment method for lowering the FFA to the acceptable level of 3% for Biodiesel production using homogenous base catalyzed transesterification.

BACKGROUND OF CHEMICAL RE-ESTERIFICATION

The chemical re-esterification process is one of the old high FFA pre-treatment methods for food grade products. It has been in existence for more than centuries [39]. It converts the free fatty acid into neutral glycerides by reesterification with the free hydroxyl groups remaining in the oil (or with added hydroxyl groups from glycerol) at a high temperature, with or without catalyst [38]. The reaction starts with the formation of monoglycerides, which is further esterified to diglycerides and then to a triglyceride [40]. Contrary to loss of oils during pretreatment by other FFA pre-treatment processes, the reesterification increases the yield of neutral oil. The water formed during reaction lead to the establishment of equilibrium between the reactants under the experimental conditions and it should therefore be removed. Several approaches have been proposed to remove water in the reaction mixture. The use of an inert gas or air and to maintain vacuum have been suggested to eliminate water from the reaction mixture [39].

FACTORS AFFECTING THE CHEMICAL RE-ESTERIFICATION REACTION

The reaction temperature, amount and type of catalyst, and amount of glycerol are the main factors that are said to affect the yield of the chemical re-esterification process in converting free fatty acid into triglycerides.

Effect of Temperature on Chemical Re-Esterification

Literatures show that chemical re-esterification can occur at different temperatures, depending on the type oil used. Variable temperatures of 180°C, 220°C and 230°C were used by Felizardo et al. [41] in pre-treating acidulated soap stock of 50% FFA. It was found that temperature increase favors the reaction kinetics considerably faster at 230°C. However, more significant difference in FFA drop seems to occur when the temperature increases from 180°C to 220°C. The FFA content of the acidulated soapstocks was reduced from 50% to 5% after 3 h of reaction at 200°C. Similar temperature trend were also observed by [39,42,43] in re-esterification of high FFA rice bran oil whereby the rate at which raw rice bran oil re-esterifies was maximum between 180°C and 200°C. De and Bhattacharyya [44] show that the reaction temperature of 210°C was more effective than temperature below 200°C in re-esterifing rice bran oil containing high FFA (9.5% to 35.0%) with monoglycerides.

The reaction temperature was also found to influence the rate of re-esterification process by Ebewele, Iyayi and Hymore [45], in chemical re-esterification of high acidic rubber seed oil with 37.69% FFA. At low temperature of 150°C the FFA was lowered to about 7.03% in 6 h. While at 200°C, the FFA dropped to 1.5% over the same period. On increasing temperature further to 250°C, the reduction in FFA was fastest within the first two hours. However, the FFA dropped to 3.88% after 6 h of the reaction time. It is supposed that there was a small degree of fat splitting at this elevated temperature after being held for 6 h. The rate at which FFA re-esterifies was at its maximum between 200°C and 250°C.

Effect of Amount and Type of Catalyst

The chemical re-esterification of free fatty acid is affected by the type and amount of catalyst used, although reaction can also proceed without catalyst [39]. An extensive research on different types of catalysts was done by Feuge, Kraemer and Bailey [46] whereby $AlCl_2 \cdot 6H_2O$, Al_2O_3, SnO_2, $SbCl_3$, $HgCl_2$, FeO, $NiCl_2 \cdot 6H_2O$, $NaOH$, $MgCl_2 \cdot 6H_2O$, MgO, $MnCl_2 \cdot 4H_2O$, $PbCl_2$, ZnO, $FeCl_3 \cdot 6H_2O$, $CdCl_2 \cdot 2.5H_2O$, PbO, MnO_2, $ZnCl_2$, $SnCl_2 \cdot 2H_2O$, $SnCl_4 \cdot 5H_2O$ and HCl were tried by chemical reesterification of the mixed fatty acids obtained by saponification of peanut oil with 90.3% FFA under reduced pressure (20 mmHg) and at the temperature of 200°C. Only $SnCl_2 \cdot 2H_2O$, $SnCl_4 \cdot 5H_2O$ and $ZnCl_2$ were found to be excellent in catalytic activity and the FFA of oil drop from 90.3% to 2.8%, 2.4% and 3.5% respectively for 6 h. When the reaction was uncatalyzed, the FFA drops to 5.34 after 8 h and at an elevated temperature of 241°C.

The rate of re-esterification was observed to be slow in the absence of catalyst by Ebewele, Iyayi and Hymore [45]. In this reaction, FFA was reduced to 15.38% from 37.69% in 6 h without catalyst. However, on using zinc dust (0.25% by weight of oil) and zinc chloride (0.15% by weight of oil) significant reduction in FFA was achieved. Zinc dust lowered the FFA of rubber seed oil from 37.69% to 1.50% while Zinc chloride lowered the FFA to about 1.27% within 6 h of reaction time. No significant reduction of FFA was observed when the two catalysts were combined.

Felizardo et al. [41] tried metallic zinc and dehydrated zinc acetate as chemical re esterification catalysts. The catalysts concentrations used were 0.1%, 0.2% and 0.3%w/w of the acidulated soap stock. Both metallic zinc and dehydrated zinc acetate catalysts showed almost the same effect on reaction kinetics. On increasing catalyst dose the reaction kinetic was also increasing until a reaction time of 1 h, however the final acidity did not seem to be affected after 1 h. It was also shown that 2 h will be required to achieve the same drop of FFA without catalyst. The effect of the catalyst dose on re-esterification of rice bran oil with 50% and 70% excess glycerol was also investigated by Singh and Singh [43] with 7 h of reaction time. $SnCl_2$ catalyst concentrations of 0.1%, 0.15%, 0.2%, 0.25% and 0.3%w/w were used. At the stated conditions, 0.2%w/w catalyst concentration was found to be optimum in lowering the acid value of the rice bran oil in both 50% and 70% excess of glycerol. Bhattacharyya and Bhattacharyya [42] investigated the effect of two catalysts namely stannous chloride and an aromatic sulphonic acid (p-toluene sulphonic acid) on the extent of re-esterification of FFA in rice bran oil with added glycerol. The catalysts were shown to influence the re-esterifi- cation rate only during the initial 2 h. The p-toluene sulphonic acid was found to be more effective by lowering the rice bran oil with 15% - 30% FFA to low levels (1.6% - 4.0%) by reesterification with glycerol.

Wang, et al. [47] tried the super acid solid catalyst SO_4^2 -/ZrO_2-Al_2O_3 in the chemical re-esterification before homogeneous base transesterification. The FFA in the waste cooking oil with an acid value of 88.4 mg KOH/g was lowered to 1.414 mg KOH/g. The re-esterification efficiency was found to be 98.4%. The catalyst showed good activity in catalyzing the re-esterifying waste cooking oil by glycerol. Their work also shows the advantages of easy separation of excess glycerol and less catalyst loading (0.3%w/w).

Effect of Amount of Glycerol

The effects of amount glycerol on the re-esterification reaction were studied by Felizardo et al. [41]. The experiments were performed at 220°C with a glycerol excess of 4%, 11% and 52%. The use of more than 10% (molar ratio glycerin/FFA = 1.10) excess glycerol did not show any improvements in the reaction kinetics at a temperature of 220°C.

In their study, Ebewele, Iyayi and Hymore [45], the stoichiometric amount of glycerol (4.3%w/w of oil) in re-esterifing rubber seed oil of 37.69% FFA was shown to be significant in FFA reduction as compared to when no glycerol was used in the reaction. However, using 5.6%w/w of oil that is 30% excess of glycerol there was no significant improvement in FFA reduction as compared to the stoichiometric amount of glycerol. With 30% excess of glycerol, the rate of FFA reduction was rapid during the initial 2 h of reaction and thereafter decreases considerably. This could possibly due to high reesterification reaction occurring between the hydroxyl groups from the added glycerol and FFA at the initial stage which leads to an increase in triglycerides content. A reduction in FFA from 37.69% to 1.5% was achieved in a reaction time of 6 h with 4.3% glycerol (stoichiometric amounts) at 200°C while under the same reaction condition and time the FFA dropped from 37.69% to about 15% when no glycerol was used. In this case reduction in FFA content is thought to be the reaction between FFA and the free hydroxyl groups remaining in the oil (Bhosle and Subramanian, 2005).

Bhattacharyya and Bhattacharyya [42] studied the effect of the amount glycerol on the extent of re-esterification of raw rice bran oil. The addition of glycerol was shown to increase the rate of reaction. The excess theoretical amounts of glycerol used were 10%, 30% and 50%. After 6 h of reaction, the FFA was reduced from 15.3% to 4% by using 50% excess amount of glycerol while the drop in FFA was from 15.3% to 6%, 5.6% and 4.8% for 10%, 20% and 50% excess glycerol, respectively.

Singh and Singh [43] tried to use 50%, 70% and 100% in excess of the theoretical amount of glycerol required in re-esterification of rice bran oil with acid value of 24.3 mg KOH/g. When using 50% excess glycerol, the drop in acid value was about 19.3% at 200°C for 6 h. On increasing the excess glycerol up to 70%, the re-esterification rate was faster and the maximum reduction in acid value was 20.2% at 200°C within 4 h. The use of 100% excess glycerol followed similar trend to that of 70% excess glycerol. However, the impact of increasing the amount of glycerol was not encouraging as the maximum drop in acid value was only 20% after 5 h.

APPLICABILITY OF CHEMICAL RE-ESTERIFICATION IN BIODIESEL PRODUCTION

The chemical re-esterification can lower the high FFA content in biodiesel for homogeneous base catalyzed transesterification technology as shown in Table 1. It is possible to lower the FFA of the oil to less than 3%, which is an acceptable requirement for efficient production of biodiesel with homogeneous base transesterification. Sousa, Lucena and Fernandes [48] use the glycerol by product from transesterification to re-esterify castor oil with an FFA of 2.36%. The high solubility of castor oil in glycerol due to hydroxyl group on the castor fatty acid glycerol makes it possible to lower the FFA to 0.22% for 2 h at 120°C without catalyst. Their results confirmed that glycerol (produced during the transesteri- fication reaction) can be used to re-esterify the oil before its use in the production of biodiesel. In most of the reviewed literatures, the process has been used mostly in producing edible grade products whereby sensory properties and colour are of importance and therefore limit further exploration of different catalysts which are not good for edible grade product. Therefore, more researches in understanding the kinetics, application of different catalysts and optimizing process are still required.

CONCLUSIONS

The reviewed literature proved that the chemical re-esterification can be used as a pre-treatment method for high FFA feedstock for biodiesel production. In fact, the process shows the potentiality of reducing FFA to less than 3% which is required for homogeneous base catalyzed transesterification. This process can also utilize the glycerol from the transesterification process and would therefore lower the cost of biodiesel. There is a need for further research on this area, since little is still known on the optimization of the process especially for processing biodiesel feedstock whereby sensory properties and color are not important. The chemical re-esterification can be more easily implemented than acid esterification and thereby avoids the need for neutralization and alcohol removal steps.

Table 1: The effect of chemical re-esterification on the final amount of FFA

Oil type	Time (h)	Temperature (°C)	Catalyst	Amount of excess ycerol	Initial FFA (%)	Final FFA (%)	Sources
Rice Bran oil	6	200	p-toluene sulphonic acid	50%	15.3	1.6	[42]
Rice Bran oil	6	200	p-toluene sulphonic acid	50%	20.5	3.1	
Rice Bran oil	6	200	$SnCl_2$	70%	24.3	3.0	[43]
Rice Bran oil	6	200	$SnCl_2$	0%	64.7	0.9	[39]
Rubber seed oil	6	200	$ZnCl_2$	4.3%	37.69	1.5	[45]
Mixed fatty acids obtained by saponification of peanut oil	4	200	$SnC14\cdot5H20$	Stoichiometric amount	90.3	1.8	[46]
Waste cooking oil	4	200	24SO /ZrO2-Al2O3	70%	44.42	0.707	[47]
Castor oil	2	120	No catalyst	100%	2.36%	0.22%	[48]

ACKNOWLEDGEMENTS

The authors wish to acknowledge the Universiti Sains Malaysia for laboratory work and ESEPRIT project at University of Dar es Salaam under for financial support.

REFERENCES

1.   K. Alekett, "World Oil and Gas 'Running Out'," CNN, 2003. http://edition.cnn.com/2003/WORLD/europe/10/02/global. warming/index.html.

2.   J. Van Gerpen, "Biodiesel Processing and Production," Fuel Processing Technology, Vol. 86, No. 10, 2005, pp. 1097-1107. http://dx.doi.org/10.1016/j.fuproc.2004.11.005.

3.   L. C. Meher, D. Vidya Sagar and S. N. Naik, "Technical Aspects of Biodiesel Production by Transesterification— A Review," Renewable and Sustainable Energy Reviews, Vol. 10, No. 3, 2006, pp. 248-268. http://dx.doi.org/10.1016/j.rser.2004.09.002.

4.   M. Kim, C. DiMaggio, S. Yan, S. O. Salley and K. Y. S. Ng, "The Synergistic Effect of Alcohol Mixtures on Transesterification of Soybean Oil Using Homogeneous and Heterogeneous Catalysts," Applied Catalysis A: General, Vol. 378, No. 2, 2010, pp. 134-143. http://dx.doi.org/10.1016/j.apcata.2010.02.009.

5.   J. Zhang, S. Chen, R. Yang and Y. Yan, "Biodiesel Production from Vegetable Oil Using Heterogenous Acid and Alkali Catalyst," Fuel, Vol. 89, No. 10, 2010, pp. 2939-2944.http://dx.doi. org/10.1016/j.fuel.2010.05.009.

6.   Z. Helwani, M. R. Othman, N. Aziz, J. Kim and W. J. N. Fernando, "Solid Heterogeneous Catalysts for Transesterification of Triglycerides with Methanol: A Review," Applied Catalysis A: General, Vol. 363, No. 1-2, 2009, pp. 1-10.http://dx.doi. org/10.1016/j.apcata.2009.05.021.

7.   C. Ngamcharussrivichai, P. Totarat and K. Bunyakiat, "Ca and Zn Mixed Oxide as a heterogeneous Base Catalyst for Transesterification of Palm Kernel Oil," Applied Catalysis A: General, Vol. 341, No. 1-2, 2008, pp. 77-85.http://dx.doi. org/10.1016/j.apcata.2008.02.020.

8.   A. Kawashima, K. Matsubara and K. Honda, "Development of Heterogeneous Base Catalysts for Biodiesel Production," Bioresource Technology, Vol. 99, No. 9, 2008, pp. 3439-3443. http://dx.doi.org/10.1016/j.biortech.2007.08.009.

9.   J. Boro, D. Deka and A. J. Thakur, "A Review on Solid Oxide Derived from Waste Shells as Catalyst for Biodiesel Production," Renewable and Sustainable Energy Reviews, Vol. 16, No. 1, 2012, pp. 904-910. http://dx.doi.org/10.1016/j.rser.2011.09.011.

10.   M. Zabeti, W. M. A. Wan Daud and M. K. Aroua, "Activity of Solid Catalysts for Biodiesel Production: A Review," Fuel Processing Technology, Vol. 90, No. 6, 2009, pp. 770-777.http://dx.doi.org/10.1016/j.fuproc.2009.03.010.

11.   M. Di Serio, R. Tesser, L. Pengmei and E. Santacesaria, "Heterogeneous Catalysts for Biodiesel Production," Energy & Fuels, Vol. 22, No. 1, 2008, pp. 207-217.

12.   A. P. Chouhan and A. K. Sarma, "Modern Heterogeneous Catalysts for Biodiesel Production: A Comprehensive Review," Renewable and Sustainable Energy Reviews, Vol. 15, No. 9, 2011, pp. 4378-4399. http://dx.doi.org/10.1016/j.rser.2011.07.112.

13.   Z. Helwani, M. R. Othman, N. Aziz, W. J. N. Fernando and J. Kim, "Technologies for Production of Biodiesel Focusing on Green Catalytic Techniques: A Review," Fuel Processing Technology, Vol. 90, No. 12, 2009, pp. 1502-1514.http://dx.doi.org/10.1016/j.fuproc.2009.07.016.

14.   D. Bacovsky, W. Körbitz, M. Mittelbach and M. Wörgetter, "Biodiesel Production: Technologies and European Providers," Report T39-B6, 2007.

15.   Y. Zhang, M. A. Dube, D. D. McLean and M. Kates, "Biodiesel Production from Waste Cooking Oil: 1. Process Design and Technological Assessment," Bioresource Technology, Vol. 89, No. 1, 2003, pp. 1-16. http://dx.doi.org/10.1016/S0960-8524(03)00040-3.

16.   J. Van Kasteren and A. Nisworo, "A Process Model to Estimate the Cost of Industrial Scale Biodiesel Production from Waste Cooking Oil by Supercritical Transesterification," Resources, Conservation and Recycling, Vol. 50, No. 4, 2007, pp. 442-458.http://dx.doi.org/10.1016/j.resconrec.2006.07.005.

17. J. M. Encinar, J. F. Gonzalez and A. Rodriguez-Reinares, "Biodiesel from Used Frying Oil. Variables Affecting the Yields and Characteristics of the Biodiesel," Industrial & Engineering Chemistry Research, Vol. 44, No. 15, 2005, pp. 5491-5499. http://dx.doi.org/10.1021/ie040214f.

18. A. B. Chhetri, M. S. Tango, S. M. Budge, K. C. Watts and M. R. Islam, "Non-Edible Plant Oils as New Sources for Biodiesel Production," International Journal of Molecular Sciences, Vol. 9, No. 2, 2008, p. 169. http://dx.doi.org/10.3390/ijms9020169.

19. T. Krawczyk, "Biodiesel-Alternative Fuel Makes Inroads but Hurdles Remain," Inform, Vol. 7, 1996, pp. 801-815.

20. M. M. Gui, K. T. Lee and S. Bhatia, "Feasibility of Edible Oil vs. non-Edible Oil vs. Waste Edible Oil as Biodiesel Feedstock," Energy, Vol. 33, No. 11, 2008, pp. 1646-1653.http://dx.doi.org/10.1016/j.energy.2008.06.002.

21. Y. C. Sharma and B. Singh, "Development of Biodiesel: Current Scenario," Renewable and Sustainable Energy Reviews, Vol. 13, No. 6-7, 2009, pp. 1646-1651.http://dx.doi.org/10.1016/j.rser.2008.08.009.

22. A. E. Atabani, A. S. Silitonga, I. A. Badruddin, T. M. I. Mahlia, H. H. Masjuki and S. Mekhilef, "A Comprehensive Review on Biodiesel as an Alternative Energy Resource and Its Characteristics," Renewable and Sustainable Energy Reviews, Vol. 16, No. 4, 2012, pp. 2070- 2093. http://dx.doi.org/10.1016/j.rser.2012.01.003.

23. H. J. Berchmans and S. Hirata, "Biodiesel Production from Crude Jatropha curcas L. Seed Oil with a High Content of Free Fatty Acids," Bioresource Technology, Vol. 99, No. 6, 2008, pp. 1716-1721. http://dx.doi.org/10.1016/j.biortech.2007.03.051.

24. S.-Y. No, "Inedible Vegetable Oils and Their Derivatives for Alternative Diesel Fuels in CI Engines: A Review," Renewable and Sustainable Energy Reviews, Vol. 15, No. 1, 2011, pp. 131-149. http://dx.doi.org/10.1016/j.rser.2010.08.012.

25. C. C. Enweremadu and M. M. Mbarawa, "Technical Aspects of Production and Analysis of Biodiesel from Used Cooking Oil—A Review," Renewable and Sustainable Energy Reviews, Vol. 13, No. 9, 2009, pp. 2205-2224. http://dx.doi.org/10.1016/j.rser.2009.06.007.

26. B. Freedman, E. H. Pryde and T. L. Mounts, "Variables Affecting the Yields of Fatty Esters from Transesterified Vegetable Oils," Journal of the American Oil Chemists' Society, Vol. 61, 1984, pp. 1638-1643.

27. M. Canakci and J. Van Gerpen, "Biodiesel Production from Oils and Fats with High Free Fatty Acids," 1999.

28. M. A. Hanna and F. Ma, "Biodiesel Production: A Review," Bioresource Technology, Vol. 70, 1999, pp. 1-15.

29. J. M. Marchetti, V. U. Miguel and A. F. Errazu, "Possible Methods for Biodiesel Production," Renewable and Sustainable Energy Reviews, Vol. 11, No. 6, 2007, pp. 1300- 1311. http://dx.doi.org/10.1016/j.rser.2005.08.006.

30. C. Tongurai, S. Klinpikul, C. Bunyakan and P. Kiatsimkul, "Biodiesel Production from Palm Oil," Songklanakarin Journal of Science and Technology, Vol. 23, 2001, pp. 832-841.

31. G. Knothe, J. Van Gerpen and J. Krahl, "Basics of the Transesterification Reaction," In: The Biodiesel Handbook, AOCS Press, Champaign, 2005, pp. 26-41.

32. M. Dorado, E. Ballesteros, J. De Almeida, C. Schellert, H. Löhrlein and R. Krause, "An Alkali-Catalyzed Transesterification Process for High Free Fatty Acid Waste Oils," Transactions of the ASAE, Vol. 45, 2002, pp. 525-529.

33. A. Kumar, A. Kumar Tiwari and H. Raheman, "Biodiesel Production from Jatropha Oil (Jatropha curcas) with High Free fatty Acids: An Optimized Process," Biomass and bioenergy, Vol. 31, No. 8, 2007, pp. 569-575.http://dx.doi.org/10.1016/j.biombioe.2007.03.003.

34. S. V. Ghadge and H. Raheman, "Biodiesel Production from Mahua (Madhuca indica) Oil Having High Free Fatty Acids," Biomass and Bioenergy, Vol. 28, No. 6, 2005, pp. 601-605. http://dx.doi.org/10.1016/j.biombioe.2004.11.009

35. M. Sheng, D. L. Tian and G. M. Cao, "Production of Biodiesel Fuel from Wast Edible Oil," China Academic Journals, Vol. 26, 2008.

36. D. Y. C. Leung, X. Wu and M. K. H. Leung, "A Review on Biodiesel Production Using Catalyzed Transesterification," Applied Energy, Vol. 87, No. 1, 2010, pp. 1083- 1095.http://dx.doi.org/10.1016/j.apenergy.2009.10.006.

37.   J. Van Gerpen, B. Shanks, R. Pruszko, D. Clements and G. Knothe, "Biodiesel Production Technology," Department of Energy, Washington DC, 2004.

38.   A. J. C. Anderson, "Refining of Oils and Fats for Edible Purposes," 2nd Edition, Pergamon Press, London, 1962, pp. 92-103.

39.   B. M. Bhosle and R. Subramanian, "New Approaches in Deacidification of Edible Oils—A Review," Journal of Food Engineering, Vol. 69, No. 4, 2005, pp. 481-494.http://dx.doi.org/10.1016/j.jfoodeng.2004.09.003.

40.   M. Blanco, R. Beneyto, M. Castillo and M. Porcel, "Analytical Control of an Esterification Batch Reaction between Glycerine and Fatty Acids by Near-Infrared Spectroscopy," Analytica Chimica Acta, Vol. 521, No. 2, 2004, pp. 143-148.http://dx.doi.org/10.1016/j.aca.2004.06.003.

41.   P. Felizardo, J. Machado, D. Vergueiro, M. J. N. Correia, J. P. Gomes and J. M. Bordado, "Study on the Glycerolysis Reaction of High Free Fatty Acid Oils for Use as Biodiesel Feedstock," Fuel Processing Technology, Vol. 92, No. 6, 2011, pp. 1225-1229. http://dx.doi.org/10.1016/j.fuproc.2011.01.020.

42.   A. C. Bhattacharyya and D. K. Bhattacharyya, "Deacidification of High FFA Rice Bran Oil by Reesterification and Alkali Neutralization," Journal of the American Oil Chemists' Society, Vol. 64, No. 1, 1987, pp. 128-131.

43.   S. Singh and R. P. Singh, "Deacidification of High Free Fatty Acid-Containing Rice Bran Oil by Non-Conventional Reesterification Process," Journal of Oleo Science, Vol. 58, No. 2, 2009, pp. 53-56.

44.   B. K. De and D. K. Bhattacharyya, "Deacidification of High-Acid Rice Bran Oil by Reesterification with Monoglyceride," Journal of the American Oil Chemists' Society, Vol. 76, No. 10, 1999, pp. 1243-1246.

45.   R. O. Ebewele, A. F. Iyayi and F. K. Hymore, "Deacidification of High Acidic Rubber Seed Oil by Reesterification with Glycerol," International Journal of the Physical Sciences, Vol. 5, No. 6, 2010, pp. 841-846.

46.   R. Feuge, E. Kraemer and A. Bailey, "Modification of Vegetable Oils. IV. Reesterification of Fatty Acids with Glycerol," Journal of

the American Oil Chemists' Society, Vol. 22, No. 8, 1945, pp. 202-207.

47.   Y. Wang, S. Ma, L. Wang, S. Tang, W. W. Riley and M. J. T. Reaney, "Solid Superacid Catalyzed Glycerol Esterification of Free Fatty Acids in Waste Cooking Oil for Biodiesel Production," European Journal of Lipid Science and Technology, Vol. 114, No. 3, 2012, pp. 315-324.

48.   L. L. Sousa, I. L. Lucena and F. A. N. Fernandes, "Transesterification of Castor Oil: Effect of the Acid Value and Neutralization of the Oil with Glycerol," Fuel Processing Technology, Vol. 91, No. 2, 2010, pp. 194-196. http://dx.doi.org/10.1016/j.fuproc.2009.09.016.

Chapter 7

Reactivity Investigation on Iron-Titanium Oxides for a Moving Bed Chemical Looping Combustion Implementation

Diana C. Campos, Jamal Belkouch, Mourad Hazi, and Aïssa Ould-Dris

Laboratoire de Transformation Intégrée de la Matière Renouvelable, Université de Technologie de Compiègne, Compiègne Cedex, France

ABSTRACT

Ilmenite-type natural ore which is constituted mainly of iron-titanium oxide is an interesting candidate as an oxygen carrier in chemical looping combustion (CLC) process. Its reactivity was investigated using methane as reducing gas and air as oxidizing gas. Experiments were carried out in a coupled thermogravimetric-thermo differential analyzer (TGA-DTA). When temperature increases from 700°C to 1000°C, the reaction rate increases by 50 times while the oxygen transfer capacity passes from 1.8% to 12%. TG-DT analyses showed that the overall mass loss due to ilmenite reduction reached at most 12%. It corresponds to 87% of theoretical mass loss due to the transformation of Fe_2TiO_5 into Fe and TiO_2. It is established that the reduction for the iron-titanium oxides occurs in two steps: $Fe_2TiO_5 \rightarrow FeTiO_3 \rightarrow Fe + TiO_2$. The titanium reduction from the state TiO_2 to the stage Ti_3O_5 was observed as well. This behavior is supported by XRD analysis. Subsequent oxidation of the reduced mineral led to recover the starting oxide. The stability of iron-titanium oxides was established over 35 looping cycles of oxidation-reduction, with an increase of 5% of oxygen transfer capacity and reactivity in the first 5 cycles and after that, ilmenite reactivity remained constant. At high temperatures, catalytic effect of ilmenite on methane decomposition leading to carbon deposition is observed. The deposited carbon participates in the reactivity of the oxide.

INTRODUCTION

In recent years, the increase of greenhouse gases (GHG) emissions is considered as a major environmental problem. The use of fossil fuels for energy production is one of its main causes [1]. Since environmentally friendly sources of energy, such as solar, wind, hydrogen, biomass, etc. are not totally developed to replace completely the energy generation from fossil fuels [2], the capture of generated carbon dioxide for its subsequent sequestration has become the more favorable short-term solution.

Chemical looping combustion (CLC) is one of the most promising technologies for the capture of carbon dioxide (CO_2) at low cost, with a high efficiency. This system employs a solid oxygen-carrier (OC), typically a metal oxide, instead of air as the oxygen source for combustion. The flue gas contains mainly CO_2 and water. Since it does not contain nitrogen, the CO_2 separation, which has the major cost for CO_2 capture can be avoided.

As illustrated in Figure 1, in CLC, the solid oxygencarrier is circulated in a loop between two reactors. In combustion reactor, solid oxygen carrier reacts with fuel to produce mainly carbon dioxide and water. Those compounds can be easily separated by water condensation.

The reduced oxygen carrier is then transported to the air reactor where it is oxidized with heat releases. The exhaust gas from air reactor is composed only of oxygen depleted air. The overall reactions of the oxygen carrier can be written as:

In combustion reactor:

$$C_a H_{2b} + (2a+b) Me_x O_y$$
$$\rightarrow aCO_2 + bH_2O + (2a+b) Me_x O_{y-1}$$

(1.1)

In air reactor:

$$Me_x O_{y-1} + 1/2 O_2 \rightarrow Me_x O_y$$

(1.2)

The total amount of heat resulting from reactions (1.1) and (1.2) does not change from a conventional combustion where the fuel is in direct contact with air. Consequently, CLC process has the advantage of producing a pure stream of CO_2 without additional energy for gases separation.

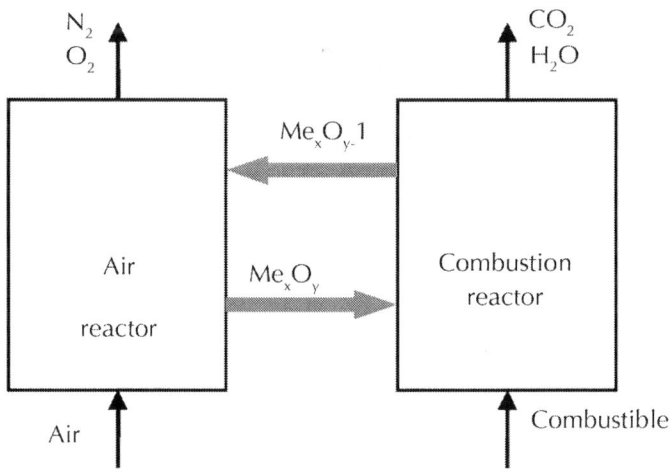

Figure 1: Reactor scheme of the CLC process.

The use of CLC was firstly introduced by Ishida [3] and developed for gaseous fuels. The earlier research on CLC focused on the selection of appropriate oxygen carriers, as one of the most critical steps. Several possible oxygen carriers have been tested: metal oxides based on nickel, copper, iron, manganese, cobalt or even mixedoxide oxygen carriers with various support materials (alumina, bentonite, rutile, etc.) [4-15]. It has been agreed that NiO and CuO were the most reactive metal oxides. However, Cu sinters at 950°C which limits the application at high temperature [5] and Ni is expensive and may generate detrimental environmental (including health and safety) effects. The Fe exhibits a moderate activity with a partial conversion of Fe^{3+}. Leion et al. [14,15] found that iron minerals are a good low cost alternative to synthetic oxygen carrier while keeping most of the suitable characteristics in CLC process. Ilmenite has proved to be the most interesting iron ores [16-18].

Ilmenite is a mineral composed of crystalline titanium-iron oxide ($FeTiO_3$). The Fe-Ti-O system including ilmenite can take several solid solutions depending on the stoichiometry between iron and titanium oxides. Nell et al. [19] have established that in oxidized ilmenite, two solid phases can be present; the first one has the M_2O_3 ilmenite structure, where $FeTiO_3$ and Fe_2O_3 are miscible within certain limits and the second is the TiO_2 rutile which is formed according to:

$$2FeTiO_3 + 1/2\,O_2 \rightarrow Fe_2O_3 + 2TiO_2 \qquad (1.3)$$

Rutile crystallites are located outside of the M_2O_3 structure and free cation sites are generated because of the addition of oxygen anionic sites. If the oxidation continues, a second phase with M_3O_5 structure appears. There is a certain miscibility reported between $FeTi_2O_5$ (Fe^{2+}) and Fe_2TiO_5(Fe^{3+}), and rutile is formed during the oxidation:

$$2FeTi_2O_5 + 1/2\,O_2 \rightarrow Fe_2TiO_5 + 3TiO_2 \qquad (1.4)$$

Besides, Briggs et al. [20] have found that pre-oxidation of mineral increases the rate reduction of samples.

Once the mineral is reduced, rutile is formed and appears outside of the iron-containing phases. During the oxidation of the reduced mineral, the original ilmenite phase is reformed and moves back into the original structure.

The industrial feasibility of CLC process for gas fuels has been successfully tested in fluidized reactor units ranging from 300 W - 500 kW using natural gas or syngas [21-26]. These reactors have the advantage of earning excellent transport properties. However, when used with iron oxide as oxygen carrier, this reactor configuration does not allow the conversion of Fe^{3+} to Fe^0 because of high partial pressure ratio CO_2/CO (see Figure 2).

According to several authors [27,28], the reduction of ilmenite to metallic iron avoids to oxidize completely the methane to the state of CO_2. Even if this inconvenience is confirmed, it is possible to take advantage of the high capacity of oxygen transfer of the ilmenite in a reactor implementation which ensures a complete conversion of methane into CO_2; for Instance, if a countercurrent (Figure 3) moving bed is used, thermodynamic equilibrium is shifted and higher conversions of iron oxides can be achieved. Thus, a higher oxygen transfer capacity takes place meaning in a significantly less amount of oxygen carrier and smaller reactor configuration. Previous work [29] in moving bed flow has demonstrated the plug flow behavior of the solids and the uniform porosity distribution into the bed. These characteristics are suitable to a good solid-gas contact and therefore a good reactivity on CLC can be expected.

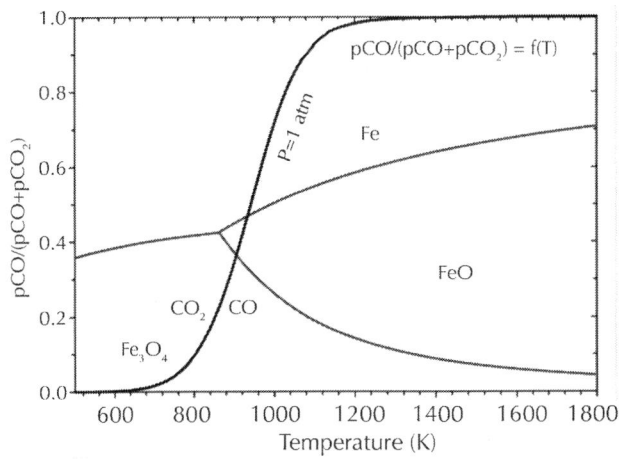

Figure 2: Thermodynamics of the Fe_2O_3 reaction with CO.

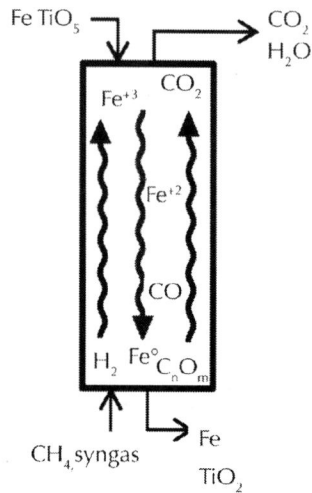

Figure 3: Countercurrent moving bed schema for using in the CLC process with Fe-Ti-O oxides.

Regarding, the reducing agents, most of available literature describes ilmenite reduction using gasification products (hydrogen, carbon monoxide), or carbon [16], [30-33] but only a limited number of studies have included the methane among these agents. Flamant et al. [34] indicated that from a thermodynamic point of view, the ilmenite reduction is more favorable using methane than using H_2 or CO as reducing agent. From a kinetic point of view, Adanez et al. [16] have shown that ilmenite has a decreasing reactivity towards gas combustibles as: $H_2 > CO > CH_4$.

Methane sources are diverse. It represents the major component of natural gas. It is also released into the primary thermal decomposition of fuel at amounts as higher as 20% for biomass fuels [34]. Moreover, the results obtained by Leion [14], Cuadrat [33-35] and Berguerand [36-38], when using solid fuels, unconverted methane is found in the exhaust gases from fuel reactor. Thereby, the comprehension of methane reactivity must be deepened into the global behavior of CLC process.

The aim of this paper is to contribute in understanding of ilmenite reduction using methane as a reducing agent. This study uses principally TG-DT analysis to evaluate the reactivity of ilmenite in order to be applied in a countercurrent moving bed. XR diffraction is used as a technique for solids characterization. Reactivity is discussed as a function of temperature and gas composition. The influence of carbon deposition and ilmenite behavior in cyclical reduction-oxidation are also investigated.

EXPERIMENTAL SECTION

Characterization of Materials

The raw ilmenite as well as reacted ilmenite were characterized by:

- X-ray diffraction using a powder X-ray diffractometer (XRD) Inel CPS 120 equipped with iron anticathode with filtered radiation λ = 1.936 Å and an ethane ionization curved detector allowing angles to be read simultaneously in the range 2θ = 5° - 120°.

- Measurement of BET specific surface area which is taken in a conventional static volume apparatus (Micromeritics ASAP 2020) operating with N_2 adsorption at −196°C. The samples were initially degassed during 3 hours at 300°C.

- Granulometric analysis established with a Mastersize X laser diffraction analyzer.

- He pycnometry using a Accupyc 1330 de Micromeretics Instruments Inc.

Reactivity Test

Raw ilmenite was pre-oxidized under air atmosphere at 1000°C for 24 hours. Reactivity experiments were performed using SETARAM 92 coupled TG-DT analyzer. Samples of 20 ± 2 mg of pre-oxidized ilmenite were placed in a platinum crucible and preheated (heating rate 20°C/min) under N_2 atmosphere to the desired reaction temperature ranging from 700°C to 1000°C. Then, it was exposed to a reducing gas (CH_4) or to an oxidizing gas (air) until the end of the reaction. Throughout the reduction/oxidation reactions, the weight and the heat flow of the sample was recorded continuously.

The gas flow rate for all periods and cycles was 30 mL/min (at 1 bar and 25°C). To determine the influence of water steam on carbon deposition, a bubbler was used to saturate the inlet reactive gas with steam. According to the selected temperature of the bubbler, the steam ratio was adjusted from 0% to 20%.

To determine the stability of ilmenite and its reactivity, successive reductions and oxidations were conducted over 32 cycles at 900°C. The solid was exposed to methane and air alternately to simulate the CLC process cycle. To avoid mixing between the oxidation and reduction gases, the reactor was flushed with nitrogen for 3 minutes after each stage of the cycle.

RESULTS AND DISCUSSION

Characterization of Ilmenite

XRD patterns of the initial ilmenite sample and pre-oxidized ilmenite (Figure 4) suggest that $FeTiO_3$ is the only crystalline phase present in the raw ilmenite. After oxidation, a mixture of two crystalline phases, ilmenite ($FeTiO_3$), pseudobrookite (Fe_2TiO_5) and TiO_2 are detected. Other characteristics of the raw ilmenite can be found in Table 1 in Section 3.5, later.

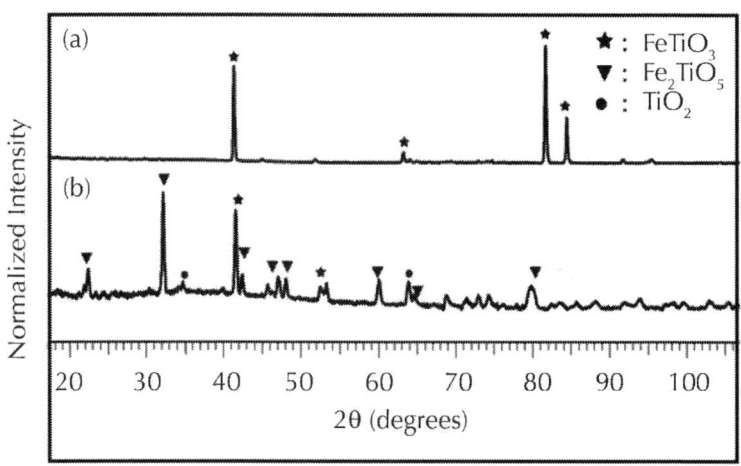

Figure 4: XRD patterns of (a) raw ilmenite and (b) preoxidized ilmenite.

Ilmenite Reactivity

Ilmenite reactivity was investigated during reduction and oxidation separately. Since the oxidized mineral is a complex mixture of Fe_2TiO_5, $FeTiO_3$ and TiO_2 (Figure 4(b)), assuming that TiO_2 is an inert material that is not affected by the reduction and the mass loss is only due to the iron cations reduction, the following reduction reactions can be expected to occur:

From Fe^{3+}: pseudobrookite

$$Fe_2TiO_5 \; (Fe_2O_3 + TiO_2)$$

$$6Fe_2TiO_5 + 6TiO_2 + 1/2\,CH_4$$
$$\rightarrow 4Fe_3Ti_3O_{10} + 1/2\,CO_2 + 2H_2O$$

$$4Fe_2TiO_5 + 4TiO_2 + CH_4 \tag{3.1}$$

$$4Fe_2TiO_5 + 4TiO_2 + CH_4$$
$$\rightarrow 8FeTiO_3 + CO_2 + 2H_2O \tag{3.2}$$

$$Fe_2TiO_5 + CH_4 \rightarrow 8Fe + 4TiO_2 + 3CO_2 + 6H_2O$$
$$\tag{3.3}$$

From Fe^{2+} Fe^{3+}: $\quad Fe_3Ti_3O_{10} \; (Fe_3O_4 + 3TiO_2)$

$$2Fe_3Ti_3O_{10} + 0.5\,CH_4 \rightarrow 6FeTiO_3 + CO_2 + H_2O \tag{3.4}$$

$$Fe_3Ti_3O_{10} \rightarrow 3Fe + 3TiO_2 + CO_2 + 2H_2O$$

(3.5)

From Fe^{2+}: Ilmenite $FeTiO_3 \left(FeO + TiO_2\right)$

$$4FeTiO_3 + CH_4 \rightarrow 4Fe + 4TiO_2 + CO_2 + 2H_2O$$

(3.6)

During the ilmenite reduction at 900°C with pure CH_4, two different rates of weight loses associated with two simultaneous endothermic DT peaks are observed (see Figure 5). The second reaction is more endothermic and faster than the first one. This behavior reveals the succession of at least two different reactions. When the weight loss reaches about 10%, the increase of the sample weight is observed.

During the ilmenite reduction, different phenomena should be taken into account. In fact, the TG signal represents the overall weight variation of the sample. It results of the signals superposition from simultaneous reactions: on one hand, the reduction of pre-oxidized ilmenite and on the other hand the deposition of carbon due to the methane decomposition. Initially, the ilmenite reduction is accompanied by the weight loss of the sample, however, when most of the available oxygen in the mineral is exhausted, the carbon resulting from the methane decomposition accumulates on the sample. In other words, when carbon deposition prevails over ilmenite reduction, the weight variation becomes positive.

Table 1: Characterization of raw and reacted ilmenite

	Pre-oxidized ilmenite	5 cycles reduced ilmenite
XRD (main species)	Fe2TiO5, TiO2, FeTiO3	T < 800°C : FeTiO3, TiO2, Fe3O4 T < 800°C : FeTiO5, TiO2, FeTiO3

True density	4575 kg/m3	4110 kg/m3
Porosity	16.07 %	36.72 %
Particle diameter	25 - 300 µm	25 - 300 µm
BET surface	0.10 m2/g	0.21 m2/g

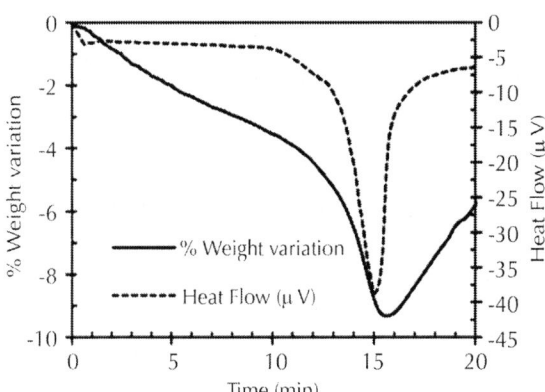

Figure 5: TG-DT analysis during pre-oxidized ilmenite reduction at 900°C.

Two explanations can justify this behavior. On one hand, knowing that the iron group (Fe, Ni, Co) is the traditional catalyst of the methane decomposition and that the iron forms exactly during the second stage of ilmenite reduction, one expect that it comes along by a carbon deposition. On the other hand, the carbon itself can contribute in the reduction of the ilmenite and thus, as long as the metallic iron is not reached, the carbon cannot accumulate on the sample. It is also possible that both described mechanisms participate simultaneously to prevent the carbon deposition during the beginning of the reduction.

The whole reaction is over in about 20 minutes. This is indicated by the coming back of the DT signal to the baseline. The weight variation obtained in thermogravimetric tests can directly be associated with the oxygen transfer capacity of the carrier (R_O) from the relationship:

$$R_O = \frac{m_{ox} - m_{red}}{m_{ox}}$$

(3.7)

where m_{ox} and m_r are respectively the mass of the most oxidized and the most reduced forms of the oxygen carrier. The oxygen transfer capacity is about 9.5% for this test (900°C). Theoretical oxygen transport capacities from total conversion of reactions (3.1) to (3.6) are summarized in Table 2. Oxidation of ilmenite was carried out using air as oxidizing agent. Results from test at 900°C are shown in Figure 6. They confirm the validity of the sequential reactions observed during the reduction test. Once again the thermogravimetric signal is overlapping the response of two phenomena: the oxidation of the reduced ilmenite (resulting in a mass decrease) and this of the deposited carbon (resulting in a mass increase).

Table 2: Theoretical oxygen transport capacity from ilmenite at different oxidation states of iron

From/To	Fe3TiO10	FeTiO3	Fe+TiO2
Fe2TiO5	1.7	5.1%	20%
Fe3Ti3O10	-	3.4%	13.6%
FeTiO3	-	-	10.5%

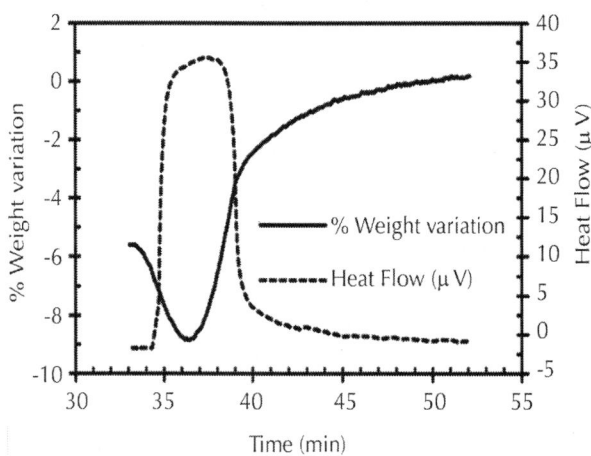

Figure 6: TG-DT analysis during ilmenite oxidation at 900°C.

The DT signal corresponds to the exothermic behavior of the overall reactions including the carbon combustion. At the end of oxidation the sample weight is slightly higher than before reduction. Thus, the R_O is enhanced over the first reduction-oxidation cycle.

Temperature Effect

The reduction behavior of pre-oxidized ilmenite was studied at 700°C, 800°C, 900°C and 1000°C. Reduced samples were characterized by XDR in order to have an idea of the nature of formed species. The temporal weight losses at these four temperatures are presented in Figure 7 and the corresponding difractogrammes in Figure 8. As expected, the reduction rates increases with increasing temperature.

The time to reach the maximum weight loss during the reduction has been decreased by 50 times between 700°C and 1000°C and R_O passed from 1.8% to 12%.

At 700°C, the maximal weight loss was reached after 250 minutes. The reduction profile did not show a significant carbon deposition after 500 min of reaction.

The XRD analysis (Figure 8(d)) revealed that the principal products of the pre-oxidized ilmenite (mostly Fe_2TiO_5) reduction are $FeTiO_3$, Fe_3O_4 and TiO_2 at this temperature. The fact that there is no change in the slope of the weight loss curve indicates a one single step reaction. Both results from TG and XRD analysis are consistent with the reactions (3.1) and (3.2), showing a partial reduction of Fe^{3+} into Fe^{2+}.

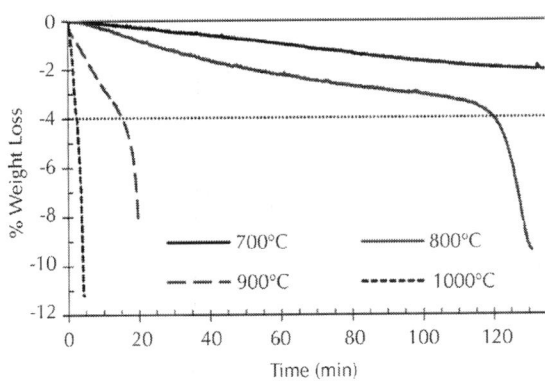

Figure 7: Effect of temperature on ilmenite reduction with pure methane (from 700°C to 1000°C).

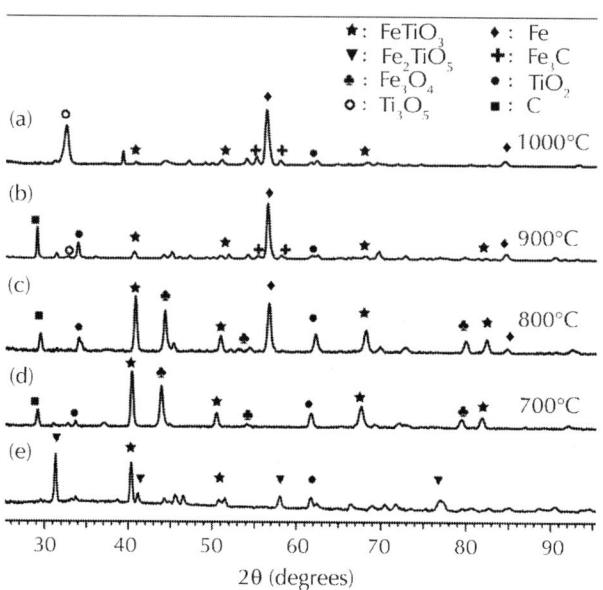

Figure 8: XRD patterns for reduced ilmenite with pure methane at different temperatures (from 700˚C at 1000˚C).

$$6Fe_2TiO_5 + 6TiO_2 + 1/2\,CH_4$$
$$\rightarrow 4Fe_3Ti_3O_{10} + 1/2\,CO_2 + 2H_2O$$
$$(R_O = 1.7\%)$$

$$(3.1)$$

$$4Fe_2TiO_5 + 4TiO_2 + CH_4$$
$$\rightarrow 8FeTiO_3 + CO_2 + 2H_2O$$
$$(R_O = 5.1\%)$$

$$(3.2)$$

At temperature higher than 800°C, reduction takes place in a two stages mechanism and a variation in the curves slopes is noted around 4% of weight loss. As presented in previous section, the first stage has a lower reduction rate than the second one.

These results are in disagreement with those reported by Abad et al. [28] indicating a higher rate for the first reduction step. The divergence could be caused by the lower methane concentration used in their work.Comparison of the measured weight loss with the theoretical values of R_O indicates that the first reaction is related to the complete reduction of Fe_2TiO_5 to $FeTiO_3$, just as in reduction at 700˚C (reaction 3.2).This result is confirmed by XRD (Figures 8(a)-(c)).

In the second stage, the metallic iron formation was proved and coupled to the disappearance of most of Fe^{2+} phases and the reaction (3.6) thereafter can be presumed. The final state of the reduction varies with the temperature and shows an incomplete reduction from Fe^{2+} to Fe^0. The higher reduction states are obtained at 1000˚C.

$$4FeTiO_3 + CH_4 \rightarrow 4Fe + 4TiO_2 + CO_2 + 2H_2O$$
$$(R_O = 10.5\%)$$

(3.6)

Besides, others phases have been detected; a graphite peak reveals the presence of deposed carbon in the overall range of studied temperatures. The migration of the carbon particles within the crystal lattice of mineral was evidenced by the apparition of the Fe_3C phase from 900˚C.

This formation was increased with temperature. At 1000˚C, the apparition of a reduced titanium oxide form (Ti_3O_5) is observed as well, besides of the loss of TiO_2 which is no longer visible. It indicates that ilmenite reactivity could not be completely approached by the only iron oxides reactivity and that mineral must be studied as such. The weight variation due to these parallel reactions is involved into the overall signal of TGA.

Effect of Carbon Deposition

As presented previously, carbon deposition is observed in all tests at temperature higher than 700˚C. Carbon deposition on the solid oxide is a fatal problem, because it lowers the mineral activity and shortens its life. Furthermore, it could diminish the CO_2 capture in the CLC process. To determine the impact of this deposition over the ilmenite reduction, steam water was added in the feed gas. Temperature was fixed at 1000°C and steam was generated by saturating the inlet courant using a bubbler.

Figure 9 shows the results from reduction test at 0% and 20% steam, it can be observed that for the first stage of reduction, the bearing of both tests is quite similar and there is no difference between the reduction rates. R_O is slightly modified by the water addition; this variation is in agreement with the diminution of deposed carbon.

In the second stage, the reduction rate decreases when the steam is added. The reduction of the reduction rate is probably due to the partial pressure decrease of methane. The whole behavior of the curves permits to affirm that over the first stage of reduction there is no carbon deposition, which is only observed when the available oxygen in the mineral decreases. The addition of water steam enables a slight higher global weight loss. At test conditions, steam water did not avoid carbon deposition completely.

The effect of carbon deposition over ilmenite during reduction was also studied. As it can be observed in Figure 5, after the solid has reached its maximal weight loss, the deposition prevails over ilmenite reduction. Figure 10 shows the behavior corresponding to the only carbon deposition phenomena (positive slope from reduction curves obtained from TGA).

It is also shown in this Figure for a comparison, a carbon deposition obtained on an inert solid (α-Alumina) instead of the ilmenite. It can be noted that the carbon deposition is favored in the presence of the reduced ilmenite, acting like a catalyst for methane cracking due to its iron content. The addition of 20% of water on reactive gas reduced the carbon deposition by a factor of 30.

Cyclical Test

Cyclical test was performed in order to investigate the behavior of ilmenite in a normal cycle of CLC process and the reactivity, deactivation and renewability of the mineral. Tests were performed over 32 cycles of reduction-oxidation with TG-TD analysis. Temperature was fixed at 900°C, air was used as oxidizing agent and pure methane was used as reducing agent. Reaction time during oxidation was set by the stabilization of the maximum weight in the case of oxidation. Reduction was considered achieved and the methane feed was stopped when the lowest mass of the sample was reached and carbon deposition begins to increase the sample mass. Figure 11 represents the sample weight variation during the first 10 cycles for pre-oxidized ilmenite.

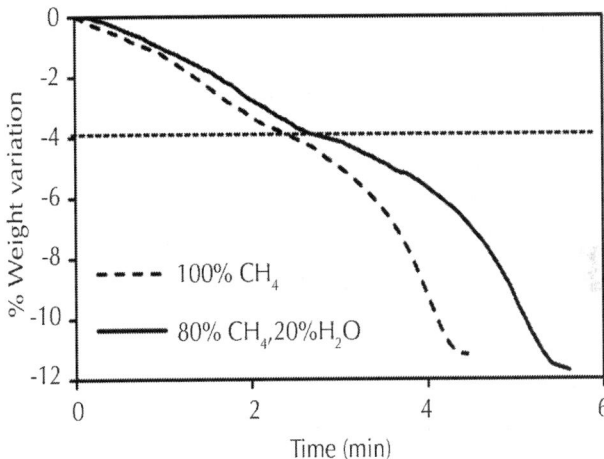

Figure 9: Effect of steam addition on ilmenite reduction at 1000°C.

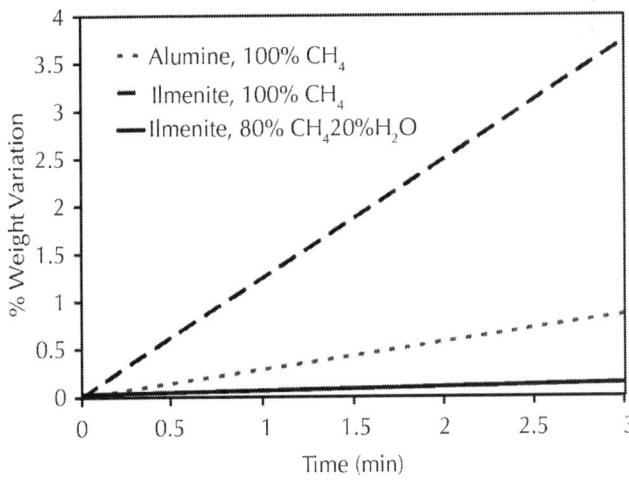

Figure 10: Weight variation due to carbon deposition over reduced ilmenite and alumina at different compositions of feed gas at 1000°C.

At first, ilmenite reacts slowly and reaches a weight variation of 9.5%. This variation is considered to be entirely due to oxygen transfer. When there was not a high amount of available oxygen in the mineral, the beginning of carbon deposition was observed and the feed of reducing gas was stopped.

After the sample is flushed with nitrogen, oxidation takes place and the TG signal has come back to initial value. There is however a phase of solid activation during the first 5 redox cycles, where both weight lost in reduction and gained in oxidation are increased. The weight loss stabilizes at an average value of 10.8% for reduction reaction and it can be also observed a growth of 0.8% of capacity of oxidation.

Both rate of reaction and oxygen transfer capacity have been improved despite the carbon deposition; this results are consistent with Cuadrat et al. experiences [39] carried out at a lower methane concentration. The improvement of ilmenite performance is associated with the porosity and reactive surface development on metallic solid. In fact, the reaction between oxygen from ilmenite and methane allows the formation of lacks in the crystalline lattice and thus enlarges the specific surface available to react. The results from characterization of reacted ilmenite exhibit a porosity development. Low values of BET surface area were measured, but a considerable increase was observed after 5 cycles of reduction-oxidation. The ilmenite properties after 32 reduction oxidation did not show significant differences compared to 5 cycles characteristics.

Figure 12 shows weight variation with time during the reduction of pre-oxidized ilmenite after different consecutive cycles. Like for the increase of oxygen capacity transfer, successive redox cycles enhance the reaction rate of ilmenite reduction. The reactivity rises around 9 times over 30 cycles.

DISCUSSION

The influence of the studied parameters (composition, temperature) on ilmenite reduction and on carbon deposition intended evaluate the ilmenite performance as an oxygen carrier in order to be used in countercurrent moving bed CLC process. A particular attention is granted to the phase of ilmenite reduction to look for the conditions allowing a high rate of reaction, a high oxygen transfer capacity and a minimization of the carbon deposit. Even if the analysis of products resulting from the reduction of the ilmenite was not realized, it is important for the CLC process to ensure that methane is mainly converted into CO_2.

The study of the complete reduction of Fe-Ti-O system until the most reduced oxidation states of iron shows the great advantage in terms of the extended oxygen transport capacity from the ore, the reducibility tests have exposed that when high methane concentrations and high temperatures are fixed, the mineral can loss of 12% of their initial weight. This value corresponds to 60% of the total amount of oxygen present in the ore as the iron oxides forms, meaning 80% of the oxygen transfer capacity when the reaction schema follows the mechanism presented above ($FeTiO_5 \rightarrow FeTiO_3 \rightarrow Fe + TiO_2$). (see Table 2). The gases analysis was not performed, but it is known from thermodynamics that the second reduction step, leads to a partial oxidation of methane to CO. This equilibrium can be shifted by using a countercurrent moving bed, and in this way a complete oxidation into CO_2 can be reached. These experiences will be the issue of future works. Test cycle demonstrate that after extent iron reductions is achieved (the case into a countercurrent moving reactor), the physicochemical structure change without losing in terms of activity of the mineral.

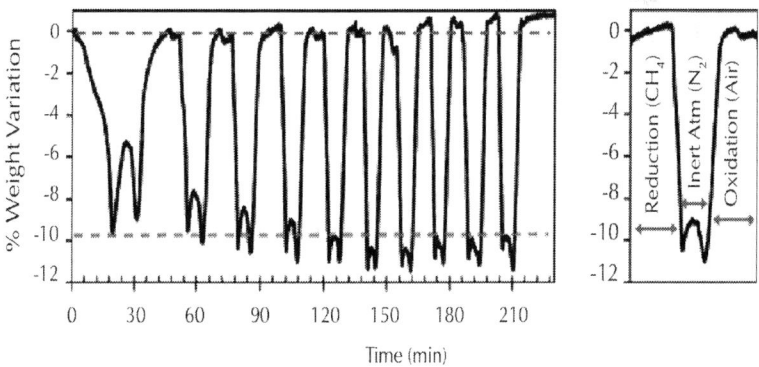

Figure 11: Weight variations during reduction-oxidation cycles using TG analysis from pre-oxidized ilmenite.

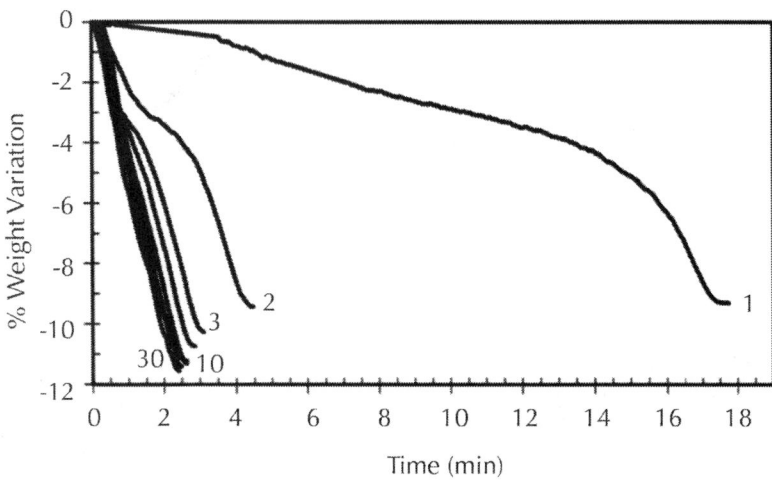

Figure 12: Weight variation in several cycles (1, 2, 3, 4, 5, 10, 20, 30) of il-menite reduction.

This study has also been established that the carbon deposition due to methane decomposition was present in all tests over 800°C. Its deposition rate was favored because of the ilmenite use. However, it has been identified that such deposition is favorable just when the oxygen capacity from the ore is exhausted.

On the other hand it has been verified that the presence of water significantly reduces this phenomenon, even in quantities as low as 20%. Anyway since water is a product of the reaction of reduction of ilmenite, this phenomenon can be avoided in the industrial realization of the process.

CONCLUSIONS

The reduction of pre-oxidized ilmenite using methane as reducing agent has been examined by TG-DT analysis. Several conclusions in connection with the chemical looping combustion can be drawn. Oxygen transfer capacity of mineral is enhanced with increasing temperature and concentration of reduction agent. The global reaction rate has increased by 50 times when temperature was increased from 700°C to 1000°C. Two stages of reduction were observed during the ilmenite reduction at temperatures higher than 800°C. These stages were confirmed by XRD. During the first stage, reduction takes place mainly as $Fe_2TiO_5 \rightarrow FeTiO_3$ and was characterized by slower rates than second one. In second stage the ilmenite reduction takes place mainly as $FeTiO_3 \rightarrow Fe + TiO_2$. At $1000°C$, a reduced titanium oxide form (Ti_3O_5) is observed simultaneously with the loss of TiO_2 which is no longer visible. It indicates that ilmenite oxygen transfer capacity in ultimate state of reduction is not only due to iron oxides reactivity but also to titanium oxide.

Ilmenite plays catalyst role in methane decomposition. However, the carbon deposition was confirmed to take place only at the end of the second reduction stage when most of oxygen available on mineral was depleted. The ilmenite renewability was confirmed over 30 cycles of reduction—oxidation, and as previously presented by Adanez et al. an enhancement on ilmenite reactivity (reaction rate and oxygen transfer capacity) has been reached after first cycles. In this test the oxygen transfer capacity rises by 2% on the first five cycles whereupon it remains constant.

ACKNOWLEDGEMENTS

This research was financially supported by the Picardie Region. The authors are gratefully acknowledged.

REFERENCES

1. IPCC, "Contribution of Working Group I to the 4th Assessment Report of the Intergovernmental Panel on Climate Change," In: S. Solomon, et al., Eds., Climate Change 2007: The Physical Science Basis, Cambridge University Press, Cambridge, 2007, p. 996.

2. I. Dincer, "Environmental Impacts of Energy," Energy Police, Vol. 27, No. 14, 1999, pp. 845-854. doi: 10.1016/j.bbr.2011.03.031.

3. M. Ishida, D. Zheng and T. Akehata, "Evaluation of a Chemical-Looping-Combustion Power-Generation System by Graphic Exergy Analysis," Energy, Vol. 12, No. 2, 1987, pp. 147-154. doi:10.1016/0360-5442(87)90119-8.

4. J. Adanez and L. F. de Diego, "Selection of Oxygen Carriers for Chemical-Looping Combustion," Energy & Fuels, Vol. 18, No. 3, 2004, pp. 371-377. doi: 10.1021/ef0301452.

5.   P. Gayan, L. Dediego, F. Garcialabiano, J. Adanez, A. Abad and C. Dueso, "Effect of Support on Reactivity and Selectivity of Ni-Based Oxygen Carriers for ChemicalLooping Combustion," Fuel, Vol. 87, No. 12, 2008, pp. 2641-2650.doi:10.1016/j. fuel.2008.02.016.

6.   K. Sedor, M. Hossain and H. Delasa, "Reactivity and Stability of Ni/Al$_2$O$_3$ Oxygen Carrier for Chemical-Looping Combustion (CLC)," Chemical Engineering Science, Vol. 63, No. 11, 2008, pp. 2994-3007. doi:10.1016/j.ces.2008.02.021.

7.   M. Johansson, T. Mattisson and A. Lyngfelt, "Investigation of Fe$_2$O$_3$ with MgAl$_2$O$_4$ for Chemical-Looping Combustion," Industrial & Engineering Chemistry Research, Vol. 43, No. 22, 2004, pp. 6978-6987. doi:10.1021/ie049813c.

8.   H.-B. Zhao, L.-M. Liu, D. Xu, C.-G. Zheng, G.-J. Liu and L.-L. Jiang, "NiO/NiAl$_2$O$_4$Oxygen Carriers Prepared by Sol-Gel for Chemical-Looping Combustion Fueled by Gas," Fuel, Vol. 36, No. 3, 2008, pp. 261-266. doi:10.1016/S1872-5813(08)60020-1.

9.   M. Johansson, T. Mattisson and A. Lyngfelt, "Creating a Synergy Effect by Using Mixed Oxides of Ironand Nickel Oxides in the Combustion of Methane in a Chemical-Looping Combustion Reactor," Energy, Vol. 56, No. 4, 2006, pp. 2399-2407. doi:10.1021/ef060068l.

10.   M. Rydén, A. Lyngfelt, T. Mattisson, D. Chen, A. Holmen and E. Bjørgum, "Novel Oxygen-Carrier Materials for Chemical-Looping Combustion and Chemical-Looping Reforming; La$_x$Sr$_{1-x}$Fe$_y$Co$_{1-y}$O$_{3-\delta}$ Perovskites and Mixed-Metal Oxides of NiO, Fe$_2$O$_3$ and Mn$_3$O$_4$," International Journal of Greenhouse Gas Control, Vol. 2, No. 1, 2008, pp. 21-36.doi:10.1016/S1750-5836(07)00107-7.

11.   E. Jerndal, T. Mattisson and A. Lyngfelt, "Investigation of Different NiO/NiAl$_2$O$_4$ Particles as Oxygen Carriers for Chemical-Looping Combustion," Energy, Vol. 94, No. 10, 2009, pp. 665-676. doi:10.1021/ef8006596.

12.   A. Abad, J. Adanez, F. Garcialabiano, L. Dediego, P. Gayan and J. Celaya, "Mapping of the Range of Operational Conditions for Cu-, Fe-, and Ni-Based Oxygen Carriers in Chemical-Looping Combustion," Chemical Engineering Science, Vol. 62, No. 1-2, 2007, pp. 533-549. doi:10.1016/j.ces.2006.09.019.

13. E. Jerndal, T. Mattisson and A. Lyngfelt, "Thermal Analysis of Chemical-Looping Combustion," Chemical Engineering Research and Design, Vol. 84, No. 9, 2006, pp. 795-806. doi:10.1205/cherd05020.

14. H. Leion, A. Lyngfelt, M. Johansson, E. Jerndal and T. Mattisson, "The Use of Ilmenite as an Oxygen Carrier in Chemical-Looping Combustion," Chemical Engineering Research and Design, Vol. 86, No. 9, 2008, pp. 1017- 1026. doi:10.1016/j.cherd.2008.03.019.

15. H. Leion, T. Mattisson and A. Lyngfelt, "Use of Ores and Industrial Products as Oxygen Carriers in ChemicalLooping Combustion," Energy & Fuels, Vol. 23, No. 4, 2009, pp. 2307-2315. doi:10.1021/ef8008629.

16. J. Adánez, A. Cuadrat, A. Abad, P. Gayán, L. F. de Diego and F. García-Labiano, "Ilmenite Activation during Consecutive Redox Cycles in Chemical-Looping Combustion," Energy & Fuels, Vol. 24, No. 2, 2010, pp. 1402- 1413. doi:10.1021/ef900856d.

17. A. R. Bidwe, F. Mayer, C. Hawthorne, A. Charitos, A. Schuster and G. Scheffknecht, "Use of Ilmenite as an Oxygen Carrier in Chemical Looping Combustion-Batch and Continuous Dual Fluidized Bed Investigation," Energy Procedia, Vol. 4, 2011, pp. 433-440. doi:10.1016/j.egypro.2011.01.072.

18. M. M. Azis, E. Jerndal, H. Leion, T. Mattisson and A. Lyngfelt, "On the Evaluation of Synthetic and Natural Ilmenite Using Syngas as Fuel in Chemical-Looping Combustion (CLC)," Chemical Engineering Research and Design, Vol. 88, No. 11, 2010, pp. 1505-1514. doi:10.1016/j.cherd.2010.03.006.

19. J. Nell, "An Overview of the Phase-Chemistry Involved in Theproduction of High-Titanium Slag from Ilmenite Feedstock," Journal of the South African Institute of Mining and Metallurgy, Vol. 100, No. 1, 2000, pp. 35-44.

20. R. A. Briggs and A. Sacco, "The Oxidation of Ilmenite and Its Relationship to the $FeO-Fe_2O_3-TiO_2$ Phase Diagram at 1073 and 1140 K," Vol. 24, No. 6, 1993, pp. 1257- 1264. doi:10.1007/BF02668194.

21. M. Johansson, T. Mattisson and A. Lyngfelt, "Use of $NiO/NiAl_2O_4$ Particles in a 10 kW Chemical-Looping Combustor," Industrial

& Engineering Chemistry Research, Vol. 45, No. 17, 2006, pp. 5911-5919. doi:10.1021/ie060232s.

22. P. Kolbitsch, J. Bolhàr-Nordenkampf, T. Pröll and H. Hofbauer, "Operating Experience with Chemical Looping Combustion in a 120 kW Dual Circulating Fluidized Bed (DCFB) Unit," International Journal of Greenhouse Gas Control, Vol. 4, No. 2, 2010, pp. 180-185.doi:10.1016/j.ijggc.2009.09.014.

23. J. Bolhàr-Nordenkampf, T. Pröll, P. Kolbitsch and H. Hofbauer, "Performance of a NiO-Based Oxygen Carrier for Chemical Looping Combustion and Reforming in a 120 kW Unit," Energy Procedia, Vol. 1, No. 1, 2009, pp. 19-25. doi:10.1016/j. egypro.2009.01.005.

24. C. Linderholm, A. Abad, T. Mattisson and A. Lyngfelt, "160 h of Chemical-Looping Combustion in a 10 kW Reactor System with a NiO-Based Oxygen Carrier," International Journal of Greenhouse Gas Control, Vol. 2, No. 4, 2008, pp. 520-530.doi:10.1016/j. ijggc.2008.02.006.

25. P. Kolbitsch, T. Proll and H. Hofbauer, "Modeling of a 120 kW Chemical Looping Combustion Reactor System Using a Ni-Based Oxygen Carrier," Chemical Engineering Science, Vol. 64, No. 1, 2009, pp. 99-108. doi:10.1016/j.ces.2008.09.014.

26. C. Linderholm, T. Mattisson and A. Lyngfelt, "LongTerm Integrity Testing of Spray-Dried Particles in a 10 kW Chemical-Looping Combustor Using Natural Gas as Fuel," Fuel, Vol. 88, No. 11, 2009, pp. 2083-2096. doi:10.1016/j.fuel.2008.12.018

27. A. Cuadrat, A. Abad, J. Adánez, L. D. Diego, F. GarcíaLabiano and P. Gayán, "Behaviour of Ilmenite as Oxygen Carrier in Chemical-Looping Combustion," Fuel Processing Technologie, Vol. 94, No. 1, 2012, pp. 101-112. doi:10.1016/j.fuproc.2011.10.020.

28. A. Abad, J. Adánez, A. Cuadrat, F. García-Labiano, P. Gayán and L. F. de Diego, "Kinetics of Redox Reactions of Ilmenite for Chemical-Looping Combustion," Chemical Engineering Science, Vol. 66, No. 4, 2011, pp. 689- 702. doi:10.1016/j. ces.2010.11.010.

29. A. Ould-Dris, Y. Molodtsof and J. F. Large, "A Classification and Design Method for Moving Bed Flow in Pipes," Powder Technology, Vol. 87, No. 1, 1996, pp. 49- 57.doi:10.1016/0032-5910(96)80758-3.

30. M. L. Vries, I. E. Grey and J. D. Fitz Gerald, "Crystallographic Control in Ilmenite Reduction," Metallurgical and Materials Transactions B, Vol. 38, No. 2, 2007, pp. 267-277. doi:10.1007/s11663-006-9015-0.

31. C. Kucukkaragoz and R. Eric, "Solid State Reduction of a Natural Ilmenite," Minerals Engineering, Vol. 19, No. 3, 2006, pp. 334-337. doi:10.1016/j.mineng.2005.09.015.

32. P. Pourghahramani and E. Forssberg, "Effects of Mechanical Activation on the Reduction Behavior of Hematite Concentrate," International Journal of Mineral Processing, Vol. 82, No. 2, 2007, pp. 96-105. doi:10.1016/j.minpro.2006.11.003.

33. G. Flamant, D. Gauthier, M. Rivot, A. Rouanet and F. Sibieude, "Mécanismes de Réduction de L'ilménite Naturelle par le Méthane dans un Réacteur à Lit Fluidisé," Powder Technology, Vol. 51, No. 3, 1987, pp. 251-260. doi:10.1016/0032-5910(87)80026-8.

34. C. V. Stevens, "Thermochemical Processing of Biomass: Conversion into Fuels, Chemicals and Power," John Wiley & Sons, Hoboken, 2011, p. 348.

35. A. Cuadrat, A. Abad, F. García-Labiano, P. Gayán, L. F. de Diego and J. Adánez, "Ilmenite as Oxygen Carrier in a Chemical Looping Combustion System with Coal," Energy Procedia, Vol. 4, 2011, pp. 362-369. doi:10.1016/j.egypro.2011.01.063.

36. A. Cuadrat, A. Abad, F. García-Labiano, P. Gayán, L. F. de Diego and J. Adánez, "The Use of Ilmenite as OxygenCarrier in a 500 W_{th} Chemical-Looping Coal Combustion Unit," International Journal of Greenhouse Gas Control, Vol. 5, No. 6, 2011, pp. 1630-1642.doi:10.1016/j.ijggc.2011.09.010.

37. A. Cuadrat, A. Abad, F. García-Labiano, P. Gayán, L. F. de Diego and J. Adánez, "Effect of Operating Conditions in Chemical-Looping Combustion of Coal in a 500 W_{th} Unit," International Journal of Greenhouse Gas Control, Vol. 6, 2012, pp. 153-163. doi:10.1016/j.ijggc.2011.10.013.

38. N. Berguerand and A. Lyngfelt, "Design and Operation of a 10 kW_{th} Chemical-Looping Combustor for Solid Fuels—Testing with South African Coal," Fuel, Vol. 87, No. 12, 2008, pp. 2713-2726.

39.   N. Berguerand and A. Lyngfelt, "The Use of Petroleum Coke as Fuel in a 10 kW$_{th}$Chemical-Looping Combustor," International Journal of Greenhouse Gas Control, Vol. 2, No. 2, 2008, pp. 169-179.

The Uses of Passive Measurement of Acoustic Emissions from Chemical Engineering Processes

Jonathan W.R. Boyd and Julie Varley

Department of Chemical Engineering and Chemical Technology, Imperial College of Science, Technology and Medicine, Prince Consort Road, London SW72BY, UK

ABSTRACT

Acoustic measurement techniques are being developed to monitor the state of equipment and the physicochemical changes within chemical engineering processes. The advantage of acoustics is that unlike other techniques, direct contact with the process under investigation is not required so intrusion can be kept to a minimum. Most research concentrates on the *active* use of low-powered ultrasound, i.e. the generation of an acoustic wave and the measurement of any changes in the wave. This review highlights the use of *passive* measurement of acoustic emissions created by a process as a potentially non-invasive, real-time monitoring technique to be used in process control. Research into the acoustic emission sources found in several chemical engineering processes such as gas–liquid mixing, solid systems and chemical reactions and the process information that can be obtained from them is discussed.

INTRODUCTION

Within the various process industries (chemical, biochemical, food, etc.) there is a need to develop accurate and reliable sensor devices to determine the physical and chemical state of a process. Sensor development is necessary to improve process design, scale-up, process control and safety. Ideally these measurement devices should be applicable to a wide range of process conditions, low costing, reliable and non-intrusive to the process being monitored. Measurements also have to be in real-time and on-line if they are to be of value in a control system.

Acoustic measurement is potentially a very useful monitoring tool. Direct contact with a process is not required allowing real-time, on-line monitoring with little or no intrusion. There are two methods of monitoring a process acoustically:

- active acoustics — the measurement of the effect a process has on a transmitted acoustic wave (usually low-powered ultrasound);
- passive acoustics — the measurement of the acoustic emission (AE) created by the process itself.

Most acoustics research has been directed towards the development of active acoustic monitoring of the physicochemical condition of processes. An acoustic wave or pulse is transmitted into the process under investigation and changes in the attenuation and speed of sound are measured to illicit process information. These active acoustic measurement techniques have been well reviewed previously (e.g. McClements, 1997). This review is intended to highlight research into the potential application of the *passive* measurement of AE (sound, ultrasound and vibration) to monitor and control processes. Our sense of hearing is used in everyday life to monitor the many events occurring around us and similarly acoustic emissions from processes could be used to monitor their state. AE are caused by physical and chemical events occurring within processes and could provide the 'listener' with valuable information as to what is occurring. Difficulties in extracting the information from the acoustic signal have been one possible reason why process AE monitoring is not more widespread.

This article introduces briefly acoustic waves, their characterisation and the basic equipment required for their passive measurement. Examples are then given describing research into the various sources of AE in chemical engineering process systems. Most AE research has concentrated on fault detection (e.g. pipe leakage, bearing failure, etc.) and has been well reviewed elsewhere (e.g. Sundt, 1979; Fowler, 1992). This review therefore describes research of AE as a means of monitoring physicochemical changes within processes such as gas–liquid mixing, powder flow and mixing and chemical reactions. Fault detection has also been briefly included to indicate other possible sources of AE within a process plant, which can be usefully accounted for. The advantages and disadvantages of the passive AE measurement technique as a monitoring technique are assessed.

ACOUSTIC WAVES, THEIR MEASUREMENT AND ANALYSIS

Acoustic Waves

Acoustics may be defined as the generation, transmission and reception of energy in the form of vibrational waves in matter (Kinsler, 1982) and is generally regarded as the study of sound and vibrations. Acoustic waves through gases, liquids and solids are longitudinal, i.e. the energy of the wave is propagated by particle vibration parallel to the direction of the wave, producing a series of high-density, high-pressure regions (compressions) and low-density, low-pressure regions (rarefractions). In solids, acoustic waves can also be transverse where the direction of particle vibration is perpendicular to the direction of the wave e.g. vibration of a string or metal rod. The low viscosity of liquids and gases means that transverse waves (or shear waves) do not occur in these phases.

Whether longitudinal or transverse an acoustic wave can be characterised by its speed, c, frequency, f, and wavelength, , which are related to each other by

$$c = \lambda f.$$

(1)

The phase speed of an acoustic wave is a characteristic property of a material and is related to the bulk modulus (elasticity) of the material B and density of the material   according to

$$c = \sqrt{\frac{B}{\rho}},$$

(2)

where B is the adiabatic bulk modulus of the liquid. In a solid, B is replaced in Eq. (2) by E, Young's modulus of elasticity of the material. For a gas, the speed of sound is

$$c = \sqrt{\frac{\gamma P_0}{\rho_g}},$$

(3)

where   is the ratio of specific heats, P_0 is the mean pressure and ρ_0 is the mean density of the gas.

The speed of sound through a material is a useful property to measure, as it will vary with the physical state of the material. Several active acoustic measurement techniques monitor the speed of sound through a material to follow the state of a process, for example, polymer curing (Shepard & Smith, 1997).

The frequency of the wave is the reciprocal of the time required for one complete cycle of a sinusoidal tone (see Fig. 1a). Sound waves are acoustic waves audible to the human ear and these occur at frequencies between 20 and 20,000Hz. Acoustics includes waves at frequencies outside the audible range. Infrasonic waves occur at frequencies below 20Hz and ultrasonic waves occur at frequencies greater than 20,000Hz.

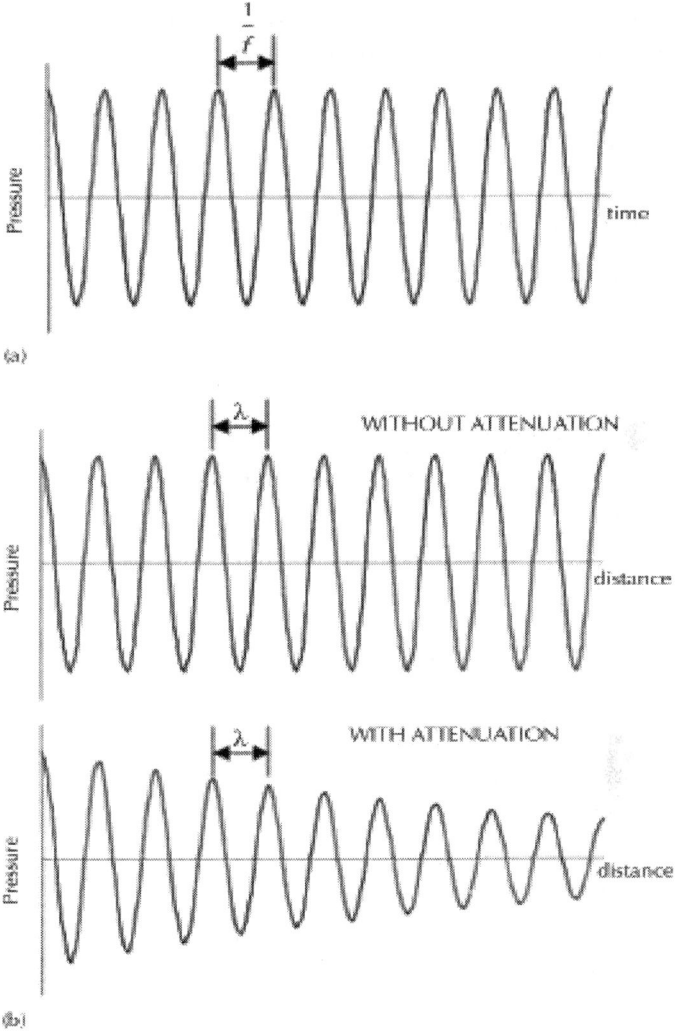

Figure 1: Illustration of pressure variation due to acoustic waves. (a) Pressure fluctuations due at a fixed distance from a continuous sound source; (b) Sound wave propagation from a sound source with and without attenuation.

As well as frequency and the speed at which it travels, an acoustic wave can also be characterised by its amplitude A. The amplitude of an acoustic wave varies with time and depends on the distance from the sound source at which it is measured. As an acoustic wave travels through a medium its amplitude will decrease due to attenuation.

Attenuation of the wave is caused by adsorption (conversion of acoustic energy to other forms of energy, mainly heat) and scattering. Scattering occurs in heterogeneous media; where the acoustic wave is incident on a discontinuity, the wave is scattered in directions other than that of the incident wave. The amount of scattering will depend on the size of the discontinuity and the frequency of the incident acoustic wave. Fig. 1b compares how the amplitude of an acoustic wave varies with distance from a source with and without attenuation. Sometimes, due to a large dynamic range of acoustic pressure, it is more convenient to describe the amplitude of the acoustic wave with a logarithmic scale. The pressure level, Lp, in decibels is defined as

$$L_p \text{ (dB)} = 20\log\left(\frac{P}{p_{ref}}\right).$$

(4)

The sound pressure, P, is compared to a reference sound pressure, p_{ref}, and the pressure level is said to be Lp decibels greater or less than the reference sound pressure. In air p_{ref} is taken as the threshold level for human hearing, 20 µPa and in water p_{ref} is usually 1 µPa. The reference pressure should be quoted when using the decibel scale.

Acoustic Measurement Equipment

In Section 3 several examples of investigations measuring the passive AE of various aspects of chemical engineering are given. Although the processes involved in these examples are quite different, the basic equipment required for the measurements is the same for all the examples. A schematic diagram representing the three basic requirements for a passive acoustic measurement system is shown in Fig. 2. Firstly the acoustic pressure fluctuations have to be detected by a sensor. The sensor signal is then usually amplified and unwanted frequencies in the signal are removed using filters. The amplified signal is then converted from analogue to digital and displayed using an oscilloscope, spectrum analyser or PC with suitable hardware and software.

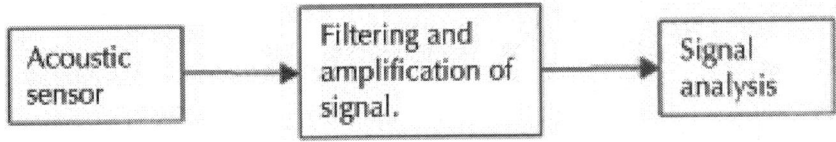

Figure 2: Standard equipment set-up for passive acoustic measurement.

An acoustic sensor must convert acoustic pressure fluctuations to an electrical current. The type of acoustic sensor most commonly used in the examples described in Section 3, contains piezoelectric crystals, which produce small electrical voltages when acted upon by pressure fluctuations. For the sensor to be calibrated, the conversion factor between voltage produced and pressure must be known. Also acoustic sensors do not respond uniformly to all frequencies, which must be accounted for in the calibration. Sensors generally have a range over which their response is regarded as uniform. In extreme conditions such as high temperatures waveguides may be required to carry the acoustic wave from the sound source to the pressure transducer (e.g. Bouchard, Payne, & Szyszko, 1994). Piezoelectric transducers, their use and waveguides have been described in further detail by Asher (1997).

Audio measurements in air (frequencies 20Hz–20kHz are made using microphones. Piezoelectric crystals are used in some microphones to convert pressure fluctuations into a voltage but condenser microphones use a capacitor effect where a thin membrane is exposed to the sound field and the movement of the membrane alters the spacing of a parallel plate capacitor. The capacitance changes result in the voltage changes across the capacitor (Turner & Pretlove, 1991).

Hydrophones are used for underwater acoustic measurements of frequencies up to the order of MHz. Water is approximately 1000 times denser than air so the hydrophone is exposed to much higher forces than a microphone in air. A hydrophone is waterproof and its working face is exposed directly to the liquid medium.

The signals generated by a sensor generally have to be amplified to provide a higher, more usable voltage. The preamplifier is placed close to, or sometimes inside, the sensor to reduce any electromagnetic interference. The equipment used to monitor the signal can then be placed metres away from the position of the sensor, for example in a control room. The preamplification stage can also include filters that remove unwanted frequencies in the signal. These unwanted signals might be due to mechanical and acoustic background noise that occurs at low frequencies. Filtering of frequencies that are twice the sampling frequency is also required to prevent aliasing occurring (discussed below). Electrical noise is generated by preamplifiers, which dictates the attainable sensitivity of the measurements.

The voltage signal from the acoustic sensor and preamplifier is continuous and therefore an analogue signal. This signal can be displayed on an analogue oscilloscope but for digital devices such as signal analysers and PCs the signal must be sampled and converted to a digital signal. The sampling frequency dictates the range of frequencies detected in the signal.

In order to accurately represent an analogue signal the minimum sampling frequency must be equal to or greater than twice the highest frequency component of the original signal (Shannon's sampling theorem) (Turner & Pretlove, 1991). The minimum sampling frequency is called the Nyquist frequency. Failure to apply Shannon's theorem will result in 'aliasing' where an alias or false frequency will appear in the signal at a much lower frequency. It should be made clear that Shannon's sampling theorem does not mean that the signal should be sampled at twice the rate of the highest frequency being investigated but twice that of the maximum frequency existing in the signal.

Basic Signal Analysis

The signal received by the signal analyser equipment is in the form of voltage (pressure) fluctuations with time. The simplest parameter used to characterise the signal 'strength' in the time domain is the root mean square (RMS) value defined by Eq. (5) for a continuous sound wave function, $p(t)$:

$$\bar{p}_{rms} = \sqrt{\frac{1}{t_2 - t_1} \int_{t_1}^{t_2} p(t)^2 \, dt}.$$

(5)

The energy of the sound wave is proportional to the square of the pressure so the RMS is a convenient parameter to account for the magnitude of the acoustic pressure fluctuations. The time signal can also be described by its probability density function (PDF), i.e. a histogram of the various sampled pressure values in the time signal. Standard deviation, skewness and kurtosis of the PDF have been used to characterise signals (e.g. Drahos, Cermak, Selucky & Ebner, 1987; Drahos & Cermak, 1989).

The linear relationship between signal values at two different times t and $t+\tau$ can be computed using the autocorrelation function (Newland, 1994)

$$R_{xx}(\tau) = \lim_{T \to \infty} \frac{1}{T - \tau} \int_0^{T-\tau} x(t)x(t + \tau)\,dt.$$

(6)

The autocorrelation function can be used to estimate how well future values of the signal can be predicted from a knowledge of the signal history, i.e. an estimate of how repetitive the signal is.

The time delay between two different signals $x(t)$ and $y(t)$ can be estimated using the cross-correlation function (Newland, 1994)

$$R_{xy}(\tau) = \lim_{T \to \infty} \frac{1}{T - \tau} \int_0^{T-\tau} x(t)y(t + \tau)\,dt.$$

(7)

The frequencies constituting a wave can be revealed using Fourier analysis. Fourier analysis of a sampled signal is performed using the discrete Fourier transform (DFT). The DFT can be computed rapidly from the sampled time signal using algorithms called the fast Fourier transform (FFT). The resulting spectrum can be used to identify the dominant frequencies in the signal. Fig. 3 demonstrates the effectiveness of using the FFT to identify the frequencies in a signal. The PDF of the original signal appears to show that the pressure fluctuations are due to Gaussian random fluctuations. The FFT, however, indicates that hidden in the signal is a sine wave at 500Hz. FFT analysis does not, however, provide any information as to when certain frequencies occur in the time domain.

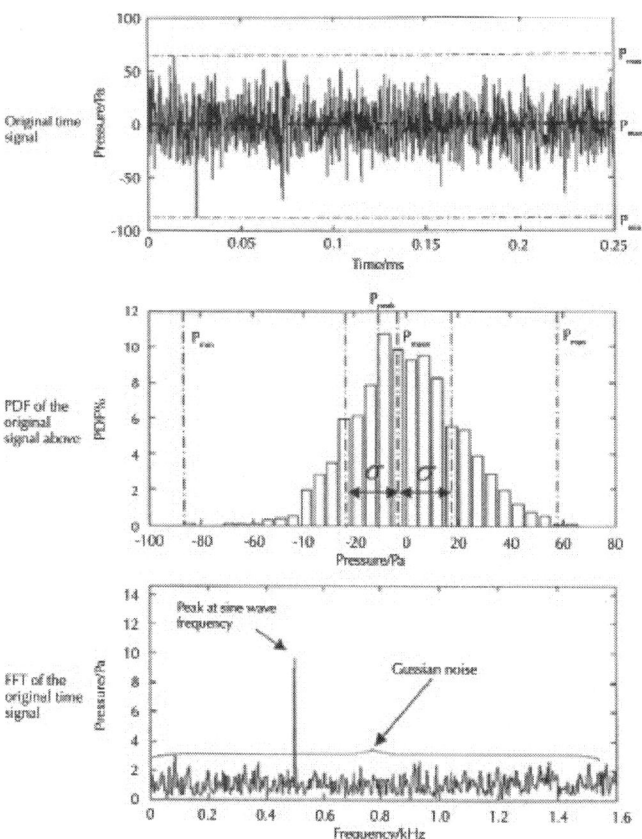

Figure 3: An example of how application of the FFT can identify important frequencies in a signal.

EXAMPLES OF PASSIVE ACOUSTIC MEASUREMENT IN CHEMICAL ENGINEERING

Monitoring of Gas–Liquid Dispersions

The mixing of gases and liquids occurs in many processes. In most cases mass transfer is required from the gas phase into the liquid (e.g.

aeration of a bioreactor) or vice versa (e.g. scrubbing of exhaust gases). In order to understand and measure mass transfer between the two phases, characteristics such as interfacial area, gas bubble size, gas hold-up, type of flow regime and the amount of power dissipated within the system are required. Existing techniques used to measure the above characteristics are generally either impractical for on-line measurements or are intrusive to the process under investigation. Probe techniques, used to measure interfacial area, bubble size and gas hold-up, require direct contact with the gas bubbles. Characterisation of the gas dispersion using photography or video requires transparent equipment and liquids, which is generally impractical for situations outside the laboratory.

Bubble Formation Sound

The sound of gas bubbles as they are formed in a liquid has been widely researched in oceanography to identify the sources of ambient sound in the sea to separate this sound from the sound of marine traffic. Bubble acoustics is well reviewed by Leighton (1994). Minnaert (1933) was the first to relate the frequency of the sound made by a bubble when formed at a nozzle to the bubble's size based on a simple energy balance calculation:

$$f = \frac{1}{2\pi r}\sqrt{\frac{3\gamma P}{\rho_l}}.$$

(8)

Eq. (8) shows a very simple inverse relationship between the frequency of AE from a bubble and its diameter (assuming a spherical bubble). For example, a 6mm air bubble in water, away from any surfaces, would oscillate at around 1000Hz. This relationship has since been investigated and demonstrated by many others (e.g. Strasberg, 1956; Leighton, Fagan, & Field, 1991). Volumetric bubble oscillation is caused by shock excitation, generally due to its formation, where the bubble's volume is displaced from its equilibrium. The bubble then oscillates producing an acoustic pressure pulse, eventually returning to its equilibrium bubble size. For small-amplitude oscillations the resulting sound pressure pulse is an exponentially damped sine wave. Large amplitude oscillations result in a non-linear pressure pulse being emitted. The magnitude of the measured sound pulse caused by bubble oscillation depends on the distance and the attenuation between the bubble source and the sensor. The presence of other bubbles can cause attenuation of the sound pulse particularly at high frequencies, which reduces the distance at which the pulse can be detected (Deane, 1997) and this must be accounted for in measurements where there is a high density of bubbles.

The simplest method of bubbling gas through a liquid is sparging gas through a single nozzle. Manasseh (1996) investigated acoustic sizing of bubbles, formed at moderate to high rates at a nozzle, as an alternative to the more usual photographic techniques. Nozzles of internal diameters 0.3–4 were used to sparge air in a box 230mm square containing water to a height of 230mm. The measured acoustic pulses close to the nozzle were the standard damped sinusoidal oscillations with some slight frequency modulation. The time traces and spectra were analysed to determine the relationship between bubble size and frequency of oscillation. Results indicated that the sound emissions from larger bubbles occur at frequencies slightly lower than predicted using Eq. (8) with values of the bubble diameter estimated from photographs. This difference between the experimental and theoretical frequencies was believed to be caused by inaccuracies in the measurement of the bubble diameter due to large shape changes. Better correlation was found if the bubble volume was estimated from the bubble formation rate in the acoustic emissions and the gas flow rate. At high gas flow rates the results diverged from the theoretical due to bubble contact and coalescence at formation.

Bubble formation processes also result in pressure fluctuations inside the body of gas in the sparger and the sparger chamber due to the volume of gas oscillating. These sparger chamber fluctuations can be detected by acoustic pressure transducers. Bubble formation, double bubbling and pairing at a point sparger have been identified from the pressure-time signal measured inside the sparger chamber (Park, Lamont Tyler, & de Nevers, 1977). Kupferberg and Jameson (1970) measured pressure fluctuations below a sieve plate to identify the causes of 'weeping' back through the sparging holes. Ruzicka et al. (1999, 2000) also measured pressure fluctuations inside the plenum containing two orifices and multiple orifices from which synchronous and asynchronous bubbling regimes were identified. Glasgow, Hua, Yiin, and Erickson (1992) used a microphone placed inside the gas space under the sieve plate of a 3 l, split column airlift reactor containing distilled water, NaCl solution and aqueous solutions of glycerol to measure acoustic spectra. Dominant sounds in the measured spectra at the sparger were ascribed to bubbles oscillating at their natural frequency when they were formed. Sound at the surface was also measured from directly above the dividing baffle. Sound from bubble disengagement and flow over the divider was found to be significant.

Estimation of the bubble size distributions in gas–liquid dispersions from the sound spectrum was investigated by Pandit, Varley, Thorpe, and Davidson (1992) for the case of a two-phase axisymmetric jet and horizontal flow in a pipe. The sound spectra measured using a Kistler 603 pressure transducer was assumed to be the result of bubbles excited into oscillation at their natural frequency by turbulent pressure fluctuations. Based on an energy balance approach, the magnitude of an individual pressure pulse emitted was estimated in terms of the displacement of the bubble radius for isothermal and adiabatic compression. The mean bubble diameters and the bubble diameter standard deviations estimated from the sound spectrum compared well with photographic results for the gas–liquid jet.

Hsi, Tay, Bukur, Tatterson, and Morrison (1985); Usry, Morrison, and Tatterson (1987); Sutter, Morrison, and Tatterson (1987) and De More, Pafford, and Tatterson (1988) measured the sound spectra in a 900mm diameter agitated vessel using a hydrophone. In all these studies, it was noted that the sound in the higher frequency range could have been caused by turbulence or the oscillations of bubbles at their natural frequency. Bubbles, visually observed in the mixing vessel, were of the right order of size to be causing the high-frequency sound but this was not confirmed quantitatively. The relationship between bubble size and the acoustic sound spectrum in agitated vessels was investigated further by Boyd and Varley 1997 and Boyd and Varley 1998. A Bruel and Kjaer hydrophone placed 10 mm from the impeller tip in a small lab-scale agitated vessel (diameter 96mm) was used to measure the acoustic emissions in the time and frequency domains (see Fig. 4). The bubble sound pulses were clearly identifiable in pressure-time samples measured close to the impeller superimposed on the hydrodynamic pressure fluctuations. These experimentally measured bubble sound pulses were characterised in terms of their magnitude, frequency and damping *for a specific set of conditions* and the results used to reasonably estimate the bubble size distribution formed at the impeller. Turbulence noise occurred at a much lower frequency range than the sound from the bubbles enabling identification of bubble sound in the spectra. Tank resonance was believed to exist but its effect on the spectra was not quantified.

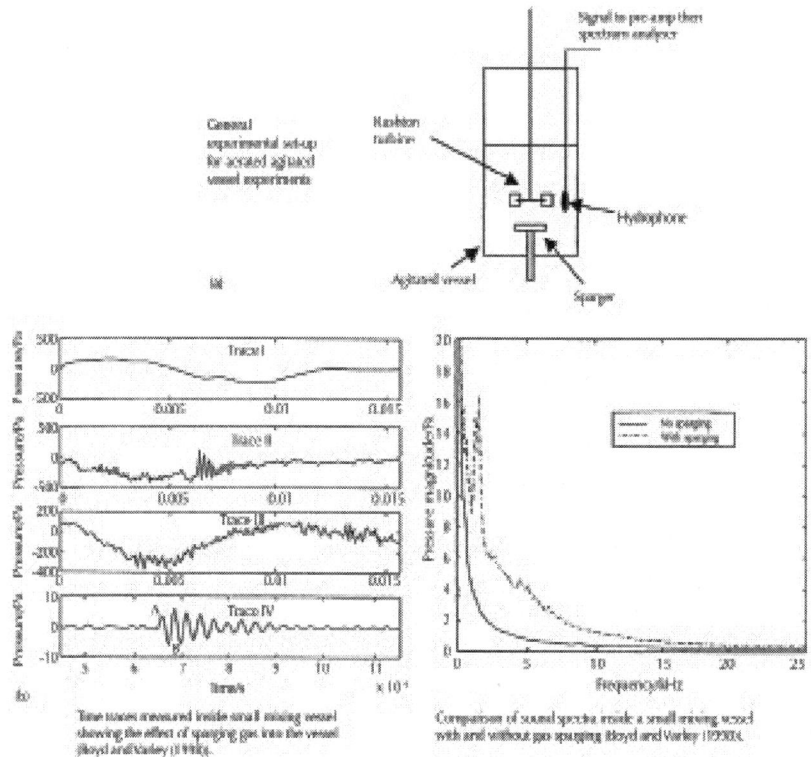

Figure 4: (a) Diagram of general equipment set-up for acoustic measurement inside an agitated vessel and (b) examples of acoustic emissions and spectra from gas mixing in agitated vessels (from Boyd and Varley (1998); reproduced with permission of the American Institute of Chemical Engineers, Copyright © 1998 AIChemE. All rights reserved).

Manasseh, Lafontaine, Davy, Sheperd, and Zhu (2000) also estimated bubble size from sound measurements inside an agitated vessel. A windowing technique was used; at a certain threshold pressure a bubble pulse was deemed to have occurred. Measurement of the signal was then activated and only the initial pulse cycles were captured in the window before calculating the spectrum. Bubble size distributions were not calculated from the generated spectra but the average bubble size was estimated from the average frequency in the spectrum. Assuming that bubbles within a critical radius of the hydrophone were detected and only one bubble existed in that region within the measurement period, the void fraction was also estimated. The void fraction estimation results were in good qualitative agreement with conductivity measurements but Manasseh et al. (2000) emphasised that these void fraction measurements could only be regarded as qualitative because the assumptions upon which the estimations were based may not be completely valid.

Bubble Bursting and Foam Break Up

Lubetkin (1989) also measured the AE due to bubble bursting at a surface. Nucleation rates of small bubbles formed at a single site beneath the surface were inferred from microphone measurements of the AE of the bursting bubbles at the surface. Rzesotarska, Rejmund, and Ranachowski (1998) monitored the RMS of the AE signals measured below short-lived static foam formed from a Triton X-100, ethanol and water mixture. The drainage process resulted in high-intensity AE. Bubble rupture was the cause of low-frequency emissions (25–70kHz) and the transport of gas through adjacent bubbles was believed to have caused emissions in the 200–300 kHz frequency band. AE activity increased with increasing ethanol concentration.

Flow Hydrodynamics in Gas–Liquid Pipe Flow

Other hydrodynamic pressure fluctuations in gas–liquid flow can be the cause of acoustic emissions. Low-frequency pressure fluctuations measured at the wall of bubbly flows have been used to characterise flow and these fluctuations can be detected as AE outside the vessel or pipework. Glasgow, Erickson, Lee, and Patel (1984) placed five pressure transducers in the wall of the upflow side of a 1360mm high, 150mm diameter airlift reactor, baffled down the middle to measure the pressure fluctuations due to the passage of gas in water and CMC solutions. Formation of large hemispherical bubbles was marked by a decaying oscillation with a frequency of approximately 20Hz, probably due to damped interfacial disturbances and/or surges in the sparger after the bubble detached. As the bubble passed the transducer in the wall there was a pressure drop at around 200ms. After the bubble had passed, there were small pressure fluctuations of around 10Hz probably due to the wake behind the bubble. Disengagement of the bubble at the top of the column was marked by decaying pressure oscillations at around 0.5Hz due to disturbances at the surface. The RMS of the pressure fluctuations increased with the gas superficial velocity.

Techniques for diagnosing the flow patterns in horizontal and vertical pipelines and bubble columns from pressure fluctuations have been reviewed by Drahos and Cermak (1989). Drahos et al. (1987) characterised horizontal two-phase flow from pressure transducer measurements in a pipe wall (inner diameter50mm, length 5m). Most pressure fluctuations occurred below 15Hz and any frequencies above that value were filtered out. Characterisation of the horizontal two-phase flow was based on parameters obtained from the signal's spectra and its probability density function. Flow regimes identified from the pressure fluctuations were bubbly flow, annular flow and slug flow. Drahos et al. (1987) stressed that the relationship between the flow regimes and the numerical parameter values were for their test system only. The investigation did illustrate that on-line identification of flow regimes in horizontal pipes was possible and removes the need for visual inspection. Process Analysis and Automation (Trade Literature) and AEA Technology (Trade literature) now market acoustic emission systems to identify flow regimes, particularly slugging, where the transducer is fixed to the outside of the pipe.

Drahos, Zahradnik, Puncochar, Fialova, and Bradka (1991) investigated pressure fluctuations within a bubble column (0.292m diameter), which contained four distribution plates. Ten pressure transducers were placed vertically in the column wall. A transducer was also used to measure the axial and radial pressure variations within the column. The standard deviations of the pressure fluctuations with time varied linearly with superficial gas velocity through the column. Correlation analysis between the time signals of two pressure transducers was used to estimate the average velocity of the recirculating streams within the vessel. Spectral analysis indicated the various characteristic frequencies of homogeneous, transition and turbulent flow at the plates within the column. An autoregressive model to describe variations in the pressure fluctuations at the transducers with time was proposed for on-line flow pattern identification.

Costigan and Whalley (1997), whilst calibrating conductivity probes to measure void fraction in a vertical pipe (8.5m tall, 32mm bore), recorded acoustic pressure transients caused by the opening and closing of a ball valve at the top of the pipe. The pressure wave propagated down the pipe and was then reflected back vertically. After a series of reflections the amplitude of the wave decayed to a value equal to the background pressure. A pressure transducer at the bottom of the column measured this pressure pulse. The speed of sound through the mixture was estimated from the frequency of the transient pulse and the vertical distance that the pulse had to travel. The speed of sound varied with the void fraction of gas in the liquid, hence measurement of the pulse frequency was used to indicate the gas void fraction in the mixture. The theoretical equation (9) was used to estimate the gas void fraction from the speed of sound (Brennen, 1995):

$$c = \left\{ [\rho_l(1 - \alpha) + \rho_g \alpha] \left[\frac{\alpha}{\gamma P_0} + \frac{(1 - \alpha)}{B} \right] \right\}^{-0.5},$$

(9)

where α is the gas void fraction, ρ_l is the density of the liquid, ρ_g is the density of the gas, γ is the polytopic index and B is the bulk modulus of the liquid. Good agreement was found between the estimated gas hold-up and the measured gas hold-up determined using conductivity measurements for void fractions between 0 and 0.5.

Flow Hydrodynamics of Gas–Liquid Flow in Packed Columns

Using acoustics as a means of diagnosing operating conditions has also been attempted for gas–liquid mixing in packed columns. Kolb, Melli, de Santos, and Scriven (1990) placed a microphone at the outlet of a small, 'almost two-dimensional' packed column; gas and liquid passed cocurrently down the column. The measured sound power spectra were used as signatures of the type of flow regime occurring within the column. In particular, the power spectrum of the pulsing regime contained a sharply defined peak at the characteristic frequency of the pulses. The peak at the pulse frequency varied systematically with liquid and gas flow rates. The pulse frequency increased with increased liquid or gas flow rates while the other was kept constant. The transition from the gas-continuous (trickling or spray) to the pulsing regime was marked by the appearance of multiple peaks in the spectrum. As the flow changed from pulsation to bubbling, the single peak in the spectrum broadens and subsides, disappearing in the bubbling regime. Acoustic measurements were used to map the flow conditions at various gas and liquid flow rates. This method of identifying the flow conditions acoustically was proposed for packed columns in opaque-walled industrial reactors where visualisation of the flow is impossible. In addition to diagnosing the operating and flow conditions, it was also suggested that acoustics could be used as a means of controlling a trickle bed reactor. The pressure fluctuations that cause AE in a horizontal packed column have been investigated byKrieg, Helwick, Dillon, and McCready (1995). Two square packed beds of inside dimensions 2.54 and 5.08cm were used, with pressure tappings at several points throughout the bed. Pressure-time traces (0–50s) and their power spectra (0–20Hz) were presented. The power spectra were dominated by peaks around 1Hz for the 5.08cm bed and 2–5Hz for the 2.54cm bed. Kolb et al. (1990) observed pulses at much higher frequencies (40–7Hz) and this was attributed to the increased scale of the thickness of the packed bed used by Krieg et al. (1995).

Flow Regimes in Agitated Vessels

The sound spectra within large-scale agitated gas–liquid dispersions have been investigated by Hsi et al. (1985), Usry et al. (1987), Sutter et al. (1987) and De More et al. (1988). In these four studies the same equipment was used: 900mm diameter vessel; 305mm diameter; 6 bladed turbine; ladder-type sparger with 3.2mm diameter orifices. 8103 Bruel and Kjaer hydrophones were used to measure the sound and the analysis was performed by a Nicolet 446 spectrum analyser.

Hsi et al. (1985) began the studies by investigating the sound spectra caused by gas dispersion at the following locations in the mixing vessel: at the sparge ring; at the impeller and in the bulk of the tank. At each position, the effect of gassing rate and the impeller speed was investigated. From the sound pressure spectra produced at the sparge ring they concluded that it was possible to distinguish between gas sparging controlled by the sparge ring itself and the natural volume pulsation frequency of the bubbles and gas sparging controlled by impeller flow. It was also concluded that a coupling phenomenon exists between the impeller and sparge ring when, under certain conditions, the sparge ring resonates at 10 times the blade passing frequency. De More et al. (1988) also measured this resonance near the impeller and it was suggested that these frequencies were within the typical shear rates found for a disc style impeller turbine. The aeration number at which maximum resonance occurred was obtained as a function of gassing rates. The impeller tip speed at maximum resonance corresponded well with the peak resonance velocity for a bubble.

Hsi et al. (1985) observed that, under various flooding and non-flooding conditions, harmonics of the impeller blade passing frequency were visible in the sound pressure spectra. As flooding approaches, these harmonics diminish. It was suggested that these harmonics were due to non-linear effects in the sound pressure waveform, although the cause of the non-linearity was not expanded on. Sutter et al. (1987) observed that these frequencies occurred at the blade passing frequency and at integer multiples of the blade passing frequency. Sutter et al. (1987) concluded that the sound spectra were directly related to the cavity formation, impeller flooding and the sparger maldistribution. Usry et al. (1987) show that the attributes of the sound pressure spectra vary in a regular way with gassing rate and impeller Re number, suggesting that this could be a method of on-line monitoring and control of the overall mass transfer coefficient.

De More et al. (1988) investigated the relationship between their cavity sound resonance measured by a hydrophone fixed to the impeller and mass transfer. The maximum cavity peak resonance was believed to correspond with the maximum production of interfacial area for electrolytic systems. Also choked flow conditions for disc style impeller were also identifiable from the sound spectra.

Acoustic Emissions from Solids and Solid–fluid Dispersions

Particle Collisions

Within processes involving the movement of solid particles, AE can be caused by particles colliding with each other or colliding with objects or vessel walls. A method of measuring particle size and size distribution in powders of rigid spheres was first described by Leach, Rubin, & Williams 1977 and Leach, Rubin, & Williams 1978a, the intention being to develop an on-line method of monitoring crushing, mixing and blending of brittle materials. When rigid particles impinge upon each other, ultrasonic AE occur. An example of the type of pulse that occurs is shown in Fig. 5a. These AE are caused by particle vibration. Several modes of vibration are possible and the frequency of each mode is related to the size and shape of the particle and their acoustic properties. Two rigid spheres of different size colliding with each other will emit acoustic signals at frequencies related to the acoustic mean of their fundamental frequency. Beats, whose frequencies depend on the difference in the fundamental frequencies of the two particles, will also occur. In reality, no two particles are exactly the same in terms of their radius and beats will always occur at frequencies orders of magnitude lower than the fundamental frequencies of the particles themselves. The acoustic measurements were made using a Bruel and Kjaer condenser microphone (type 4138) close to a rotating, foam-lined cylindrical vessel containing metallic or glass spheres of known size distributions (diameters varying between 5 µm and 3 cm). The reciprocal of the mean frequency of the beat patterns was found to increase linearly with the average diameter of the spheres (see Fig. 5b). Whenever the ratio of Young›s modulus to the particle density of different materials is the same, then the vibrational modes of the particles will also be the same. This condition applies for most brittle materials and many metals, and the technique could also be used for powder mixtures. Leach, Rubin, and Williams (1978b) demonstrated that for Gaussian particle size distributions the width of the distribution as well as the average particle size could be estimated. Leach and Rubin (1978) also demonstrated that particle size estimation from AE is relatively insensitive to the shape of the particles.

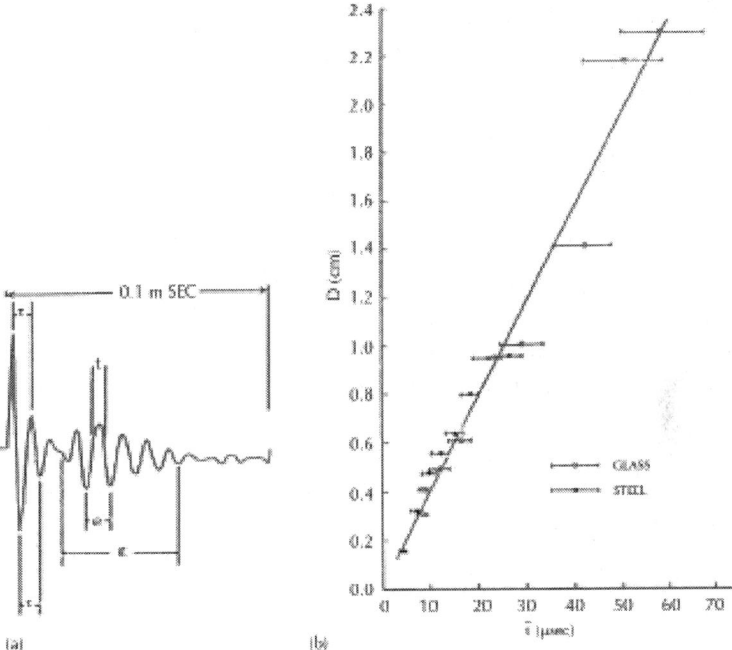

Figure 5: (a) Sound pulse due to colliding spherical steel balls in a miller and (b) an example of the relationship between average particle size and sound pulse. (Reprinted from M.F. Leach et al., Particle size distribution characterisation from acoustic emission, *Powder Technology*, *19*, 157–167, Copyright 1978, with permission from Elsevier Science).

Tily, Porada, Scruby, and Lidington (1987) simulated solid/solid mixing in an orbital mixer using a common kitchen mixer with a plastic or metal bowl. The RMS of the signal measured by a transducer attached to the bowl was followed with time for several binary solid–solid mixtures and it was found that the end point of the mixing could readily be identified when steady-state AE was reached. The amplitude of the emissions increased with increasing mass of solid and the size of the particles involved. Preliminary investigations of the mixing of solid–liquid systems showed similar variation of AE with time. Watkins, Haywood, and Scruby (1989) patented a method for monitoring the drying of wet particles or coating of solid particles with a liquid, which again follows the RMS of the AE as a means of identifying the state of the drying process. Belchamber and Collins (1989) also patented a technique for monitoring the particle size distribution in a milling process by using multivariate analysis to relate the intensity of AE from the milling process to the particle size.

Harrenstein and Brusewitz (1987), and Brusewitz and Venable (1987) proposed acoustic measurement as a means of measuring the moisture content in grains (wheat, corn, milo, soybeans) flowing from the bottom of a silo onto a pile of grain. Sound pressure level was found to increase with lower grain moisture. The variation between sound pressure level and moisture content was also dependent on the type of grain under investigation.

Bouchard et al. (1994) used an acoustic sensor to non-intrusively monitor a batch crystallisation process. An acoustic sensor was mounted on the exterior of the vessel attaching it to a structure that made good acoustical contact with the process fluid. In this case a steel rod was inserted into the vessel but alternatively a temperature probe already present in the vessel could have been used. The AE were caused by particle collisions with the steel rod. Several parameters were monitored in the acoustic signal; the most important of which were pulse count rate and total acoustic noise level. Initially, the effect of particle size on the AE was demonstrated using glass ballotini of known sizes. The pulse count rates increased with increasing amounts of material and increasing particle size. The AE from a crystallisation reaction were also monitored and neural network modelling was used to estimate the state of the crystallisation process. Fig. 6a shows that the neural network estimation of the crystallisation yield (in the form of the saturated solution temperature) provides a reasonable agreement with the actual value. It was stressed that for application of the technique in an industrial situation the neural network would have to be trained using data from the system where it is to be installed.

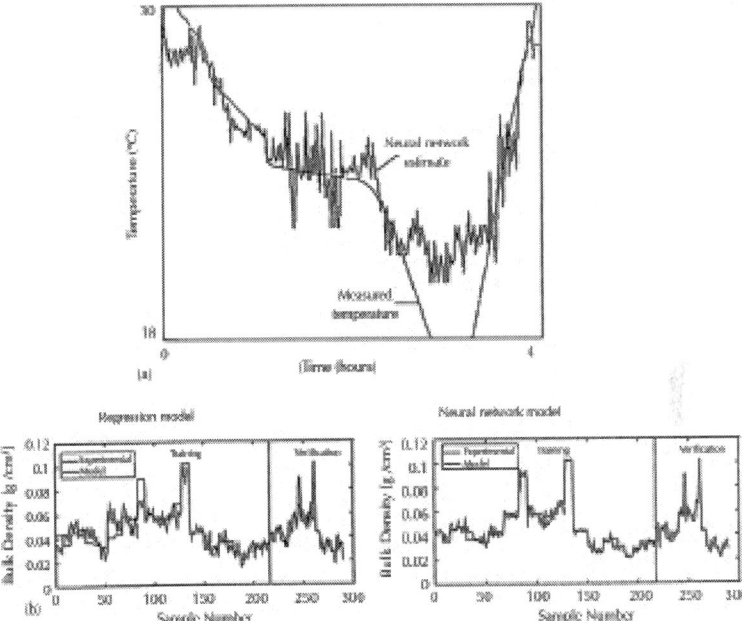

Figure 6: Examples of the use of neural networks and regression analysis of acoustic signals. (a) The use of a neural network to monitor the yield in a crystalisation process (reprinted with permission from J.G. Bouchard et al., Non-invasive measurement of process states using acoustic emission techniques coupled with advanced signal processing, *Transactions of the Institution of Chemical Engineers, 72(A)* Copyright 1994 IChemE). (b) The use of neural networks and regression software to monitor the bulk density of an extrudate (reprinted from J. Elsey et al., Acoustics based on-line estimation, *Computers and Chemical Engineering, 22,* Copyright 1998, with permission from Elsevier Science).

As described above collisions between solid particles result in AE. Folkestad and Mylvaganam (1990) measured the AE signals from the collisions of the sand particles in oil pipelines with the pipe wall. Increased concentration of sand within the oil resulted in increased AE and an unacceptable amount of sand in the line was marked by a threshold level. Fluenta market this technique for above- and underwater applications (Company literature). The advantage of using this method is that the measurement transducer is clamped onto the outside of the pipe without the need for shutting down plant operations.

Flow Hydrodynamics of Particle or Solid Flow

As for gas–liquid pipe flow discussed above, hydrodynamic pressure fluctuations can result in AE. When solid particles are either conveyed pneumatically or fluidised, hydrodynamic pressure fluctuations occur.

Pressure fluctuations causing noises in the pneumatic conveying of solids in horizontal piping were investigated by Dhodapkar and Klinzing (1993). The power spectral density function of dynamic and static pressure changes at the pipe wall was used to identify different types of flow regime. Homogeneous flow resulted in high-frequency, low-amplitude fluctuations in the power spectrum (~0–10Pa²) up to 200Hz. Stratified flow produced fluctuations near 0Hz at generally lower amplitudes (~0–3Pa²). Blowing packets created regular pressure fluctuations with marked peaks in the spectrum with amplitudes an order of magnitude greater than those for homogeneous flow (~100Pa²). Dunes in the flow created irregular fluctuations less than 4 Hz (~0–100Pa²).

The AE of flowing solid–liquid mixtures (silica flour and water) in a small diameter pipe were measured by Hou, Hunt, and Williams (1999). The peak frequencies in the measured spectra were due to the flow pulsation caused by the pumping. The magnitude of the peaks changed with the properties of the slurry. Characteristics of the flow (solid concentration, mass flow rate, volume flow rate) were related to the magnitude of important frequencies in the measured acoustic spectra using multivariate analysis.

Multivariate statistical modelling was also used by Hou, Hunt, and Williams (1998) to relate the operating parameters to the spectral characteristics of an acoustic signal measured by a transducer mounted on the top of a hydrocyclone. Power density spectra were obtained in the 0–1000Hz range. A substantial frequency range of the spectra agreed with the basic turbulence empirical law described by Eq. (10), which indicated that turbulent hydrodynamic pressure fluctuations were the cause of the emissions:

$$E(f) \alpha f^{-5/3}.$$

(10)

Peaks in the spectra were believed to be at the natural frequency of the vortex formed inside the hydrocyclone but this could not be demonstrated with the equipment available at the time. Solid concentrations, feed pressure, mass flow rate, volume flow rate and underflow concentrations were all related to statistical parameters in the time domain (maximum, minimum, mean, standard deviation, root-mean-squared value, skew and kurtosis) and the first 52 spectral characteristics of the power density spectrum (0–200Hz in the frequency domain). For all the operating parameters included in the model there was less than 3% average prediction error.

Pressure fluctuations in fluidised beds have been used to identify different types of flow regimes e.g. Brereton and Grace (1992). These pressure fluctuations can result in AE. Roy, Davidson, and Tuponogov (1990) found that pressure waves resulting from bubble formation travelled at the speed of sound and this could be estimated assuming the gas was compressed isothermally. The origins of pressure waves in fluidised beds and how they propagate were also investigated by van der Schaaf, Schouten, and van den Bleek (1998). Bubble formation and bubble coalescence resulted in compression waves that propagated upwards with amplitudes linearly dependent on the distance to the bed surface. Downward moving waves are caused by bubble eruptions at the surface of the bed, bubble coalescence and changes in bed voidage with a pressure wave amplitude independent of position in the bed. Rising gas bubbles also caused pressure fluctuations as they passed the pressure transducers with wave velocities less than $2ms^{-1}$.

Johnsson, Zijerveld, Schouten, van den Bleek, and Leckner (2000) measured dynamic pressure signals inside a fluidised bed, comparing various methods of analysing the time signal to identify flow regimes. In addition to the standard linear techniques such as those described in Section 2.3, non-linear analysis of the pressure fluctuations was also applied. Changes in the amplitude of the fluctuations gave no direct information regarding the dynamics of the system and could be caused by redistribution of the solids. Three characteristic regions in the power spectra were identified. The macro-structure of the flow occurred at frequencies up to 4Hz. Finer flow structures produced fluctuations in the 4–10Hz range that fall off in frequency as a power law or exponentially. Above 20Hz was another power law region, which had no clear dependence on the fluidisation regime.

Acoustic effects that accompany combustion in fluidised beds of inert materials have been investigated byZukowski (1999) and were proposed as a means of controlling operation of fluidised beds burning gaseous fuels. The AE from the laboratory fluidised bed at different temperatures (750–1000°C) was measured by a microphone placed outside the vessel. The AE consisted of spikes in the time signal, heard as loud 'knocks' against a background of hissing caused by explosive combustion of a certain volume of gas mixture. The magnitude of these spikes decreased in magnitude with increasing temperature of the bed because, at higher temperatures, the explosions occurred at lower positions within the bed, further away from the microphone. Above 950°C only a hissing sound remained with no spikes in the signal.

The sound emitted by a cooking extrudate as it leaves a die has been used to monitor the product quality by Elsey et al. (1998). A popping sound is caused by the bursting of steam pockets, which have formed close to the extrudate surface due to the flashing of moisture. Sound samples for 27 different operating conditions, were measured from the extruder (including noise due a motor, cutter and cooling fan), using a microphone connected to a PC. Samples of the product were collected and product qualities such as bulk density, moisture content and textural properties were measured. Tasters also gave the product a quality rating. Both time and frequency domain information was used to relate the sound produced to the operating conditions. The popping rate was obtained from the number of peaks per second in the time domain, which were more than 1.9 standard deviations from the mean value of the signal. The choice of the value of 1.9 standard deviations was the result of an optimisation procedure to give a popping rate most highly correlated with the bulk density of the product. Principal component analysis was performed on the power spectra measurements. The first 10 principal components of the frequency domain data were used for modelling purposes. Modelling was used to relate the extrudate product qualities of bulk density and fracture force, with the measured sound signal properties (popping rate, RMS power and the 10 most significant principal components). Two modelling methods were used: genetic programming (GP) and artificial neural networks (ANN) (some results are shown in Fig. 6b). Of the two models, the ANN resulted in better verification correlation coefficients for both bulk density and fracture force. The GP model, however, did produce simple correlations containing only a few parameters. Although useful to demonstrate the possibilities of the use of acoustics in an on-line control system, these derived models for the sound production at a cooking extruder are not applicable to any other extruders other than the one investigated.

Compaction

Hakanen, Laine, Jalonen, Linsaari, and Jokinen (1993) measured the AE from a roller compactor (Bepex Pharmapaktor 200/508) during the compaction of three different pharmaceutical materials (lactose monohydrate, microcrystalline cellulose and maize starch) for forces up to 60kN. The measured power spectra were divided into three bands:

- 50Hz–3.8kHz — frequency band due to the sound of machinery: the spectra in this band changed little and contributed most of the total sound power;
- 3.8–7.7kHz — very little change between conditions;
- 7.7–12.8kHz — spectra varied with changes in the compaction or compression mechanisms. In this band 'capping' (when the compressive forces exceed a limit resulting in the product being split into two and turned to yellow at its edges) was identified in the sound spectrum.

Hakanen and Laine (1995) investigated further the AE of roller compaction of microcrystalline cellulose powder and from single tablets after compaction by a single punch tablet machine. More sensitive measurement equipment was used than in the study by Hakanen et al. (1993). 'Capping' was indicated by increased AE between 17 and 23 kHz in the sound spectrum.

AE Monitoring of Chemical Reactions

Many chemical reactions, particularly those with phase changes, are acoustically active. AE monitoring has also been investigated as a means of obtaining information about specific chemical reactions. Van Ooijen, van Tooren, and Reedijk (1978) noted a loud cracking sound when $ZnCl_2$ was added to pyrazine in water (addition of pyrazine to water results in the formation of a precipitate). The cracking was heard again when the mixture was shaken hours later. Temperature increases were observed to occur at the same time as the AE. The intensity of the sound appeared to be proportional to the concentrations of both the reactants used. The sound pressure level produced by the cracking sound was analysed in terms of frequency and it was observed that the most intense part of the emissions occurred at approximately 100 kHz which is outside the range of human hearing. Van Ooijen et al. (1978) suggested that the AE were caused by the phase transition as the precipitate crystals formed, damaging the original crystals.

Betteridge, Joshlin, and Lilley (1981) expanded the study of AE caused by chemical reactions, investigating 43 reactions. The reactions occurred in a glass beaker with a piezoelectric transducer placed on the underside of the beaker (see Fig. 7a). The signal from the transducer was band passed filtered to remove frequencies outside the range of 100–300 kHz. The signal was then passed through an RMS converter and the acoustical power (see Eq. (11)) was followed with time:

$$\text{power} = \frac{(\text{RMS voltage})^2}{R},$$

$$(11)$$

where R is the resistance in parallel to the transducer output. The transducer voltage output was also measured for analysis in terms of time and frequency for certain reactions. Various methods of adding the reactants were used: direct addition, injection by syringe into the base of the beaker and peristaltic pumping into the base of the beaker. The solutions were stirred in a noise-free manner using an induction motor. Molecular sieve reactions took place in a 'crude' ion-exchange column. In this case the transducer was held against the curved surface of the column. An example of the AE caused by the addition of solid sodium hydrogen carbonate to copper sulphate solution is shown in Fig. 7b. After the initial spike caused by the addition of the solid, the acoustic energy decreased exponentially over a period of 12–13 min. Table 1 shows examples of some of the reactions and descriptions of their AE investigated by Betteridge et al. (1981) and further studies by Sawada, Gohshi, Chikako, & Furuya 1985a and Sawada, Gohshi, Chikako, & Furuya 1985b), Belchamber et al. (1986), Wentzell and Wade (1989), Cao, Wang, Wang, Lin, and Yu (1998) and Crowther, Wade, Wentzell, and Gopal (1991).

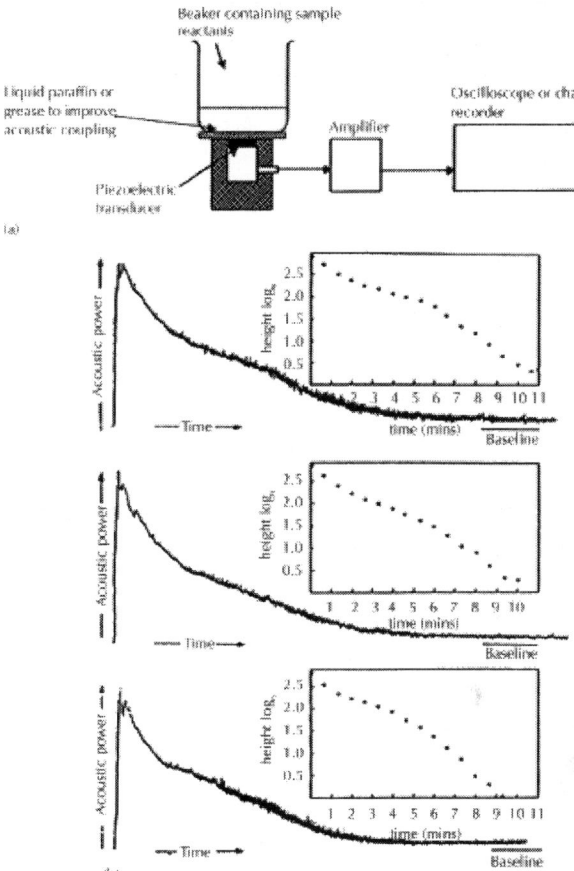

Figure 7: (a) General experimental set-up used to investigate the AE from chemical reactions. (b) Plots of acoustic energy for the reaction between 2.78g of solid sodium hydrogen carbonate and 150ml of 0.52m copper sulphate solution. (Reprinted with permission from Betteridge et al., Acoustic emissions from chemical reactions, *Analytical Chemistry, 53*, 1064–1073, Copyright 1981 American Chemical Society).

Table 1: Summary of chemical reactions investigated for acoustic emissions

Study	Reactions	Description of resulting AE
Betteridge et al. (1981)	Addition of sodium carbonate to copper sulphate solution	Initially large spike in AE power, decaying exponentially

	Addition of concentrated sulphuric acid to water or NaOH	Large spike in AE on addition of acid and rise in temperature. Spikes in AE continue with reaction. Intensity of AE decreased with increasing H_2O volume
	Oscillating reaction between iodate, peroxide and malonic acid	AE oscillated with the oscillating reaction
	Mixing organics	No AE
	Gelation reaction of 4.Msolution of calcium chloride and 2M sodium carbonate	Gradual increase in AE to max. in 3 min
		AE diminished to background after 10min but occasional bursts up to 2h after
Sawada et al. (1985a)	Addition of sodium carbonate to calcium chloride	Phase separation, precipitation and gel formation were identified from AE
Sawada et al. (1985b)	Dissolution and precipitation of sodium thiosulphate	AE from phase transitions were more intense for contraction volume changes than for
	Phase transition of p-cresol, MBBA liquid crystal and water	volume expansions
Belchamber et al. (1986)	Hydration of silica gel	Gas evolution fractured the granules causing the AE
Wentzell and Wade (1989)	Hydration of silica gel and quicklime	Granule fracture causes AE
	Aqueous dissolution of NaOH and Na_2CO_3	Effervescence AE occurred in the 50–100kHz band range
	Solid-phase transitions of C_2Cl_6	
Cao et al. (1998)	Metal-acid reactions	Significant AE in the 320–360kHz region of the power spectrum
	Carbonate/bicarbonate-acid reactions	
	Bubble effervescence	Bubble effervescence occurred below 150kHz
Crowther et al. (1991)	Electrolysis cell	AE occurred up to 800kHz. Increasing the applied voltage resulted in greater intensity of AE. Frequency mean and median, RMS and kurtosis all correlated well with the applied voltage

Betteridge et al. (1981) were the first to attempt classifying reactions by their AE using pattern recognition. The 43 reactions investigated were classified into eight clusters based on three parameters: (1) duration of the emission; (2) maximum intensity of emission; (3) a heterogeneity factor of both reactants and products (each mole of liquid in the reaction equation was given the value of zero, each mole of solid a value of 1 and each mole of gas a value of 2). Although classification of the reactions was possible, it was concluded that further work was required in the technique. It was proposed that voltage measurements in the time and frequency domain possibly could provide further parameters for understanding the processes involved in producing the AE and in classification of the reactions.

Belchamber et al. (1986) again applied pattern recognition techniques to identify gas bubble and granule fracturing AE during hydration of silica gel. Parameters that were the most ideal for classification of chemical reactions from AE were outlined by Wentzell, Lee, and Wade (1991) using Fischer weighting. Recommended descriptors were mean and median frequencies, frequency bandwidth, number of level crossings, crest factor in time and frequency domains, half-life, kurtosis and normalised percentiles of the spectrum. Parameters that depend on the magnitude of the signal, such as the RMS amplitude, were tested and produced good Fisher weights but these were not recommended for classification because they were dependent on the location of the source, which could vary from system to system.

The AE of another type of chemical reaction, the electrolysis cell, was studied by Crowther et al. (1991). The cell consisted of a stainless steel anode 120 mm in length and 7mm in diameter and a nickel cathode of the same dimensions immersed in sodium hydroxide solutions ranging from 0.1 to 2.1M. The acoustic sensor (Bruel and Kjaer broadband piezoelectric transducer, model 8312) was taped to the working electrode. The potential applied to the cell ranged from 1.4 to 5.0V. The evolution of hydrogen and oxygen bubbles at the electrodes in an electrolysis cell coincided with bursts of AE. The resulting AE were mainly ultrasonic, occurring up to 800 kHz. Some audible emissions did occur but these were filtered from the transducer signal. Acoustic power increased to a maximum at a concentration of 1.5M. Increasing the applied voltage across the cell resulted in a greater intensity of emissions in the 50–800 kHz range but approximately the same power spectrum. Descriptors of the time signal such as frequency mean and median, RMS and kurtosis all correlated well with the applied voltage.

Equipment Monitoring

The development of methods for monitoring equipment performance is important if a process plant is to run safely and reliably. Parkinson (1991) uses the term 'holistic maintenance' to describe situations in which predictive as well as the traditional preventive maintenance are used together to reduce maintenance costs and increase production performance. Measurement of AE has been investigated and used by industry as one tool to monitor equipment performance so that potential problems can be predicted and treated before they become serious. Drouillard (1988) charts the history of the development of AE monitoring for testing structural integrity. The AE monitoring of the integrity of structures, equipment parts and to detect leaks is now an established non-destructive testing method with international codes and standards.

Structural Integrity

Vahaviolos, Pollock, and Lew (1991) and Fowler (1992) review several examples where AE has been used as an indicator of structural integrity in the chemical industry by companies such as Monsanto, Dupont, Union Carbide and Exxon. AE testing is described as an economic method for 'screening', which is then followed up using further test methods such as visual inspection (Fowler, 1992). When a material undergoes deformation it releases elastic energy as AE. Transient elastic waves propagate through the material as a result of this rapid energy release. An array of resonant piezoelectric sensors, mounted on the surface of the material is used to detect these waves. When a wave 'hits' the sensor this causes the output signal from the sensor to rise above a threshold level. Various attributes of the signal are measured: number of times the signal rises above the threshold, time between the first and last crossing of the signal above the threshold and the signal strength.

Leak Detection

Leaks of a gas or a liquid are also the cause of significant AE due to resulting turbulence and/or bubble formation. Fowler (1992) describes on-line leak detection in heat exchangers in a sulphuric acid plant where there is a risk of steam at high pressure and temperature leaking and diluting the acid to cause extensive corrosion. Rather than expose the sensors to the high temperatures of the heat exchanger, wave guide probes are attached to the outside of the tube sheets and inspection ports. An alarm sounds if a leak is detected. Leaks in pipelines and their position can also be detected from the resulting AE. For example,Hunaida and Chu (1999) measured the frequency spectra of leaks in plastic pipes. The frequency content of the signal was related to the leak type, flow rate, pipe pressure and the season in which measurements were made. The attenuation rate and variation of propagation velocity with frequency were also determined. Leak localisation was also possible using cross-correlation techniques on signals upstream and downstream of the leak using the propagation velocity.

Laine, Glucina, Chamant, and Simonie (1998) used hydrophone sensors to monitor the integrity of membranes used in microfiltration and ultrafiltration to remove bacteria and cysts from water, without the need of disinfectant. Any fibre damage to the membrane could affect the treated water quality. Fibre damage (either a cut fibre or a hole in the fibre) resulted in an increase in the acoustic level of measured sound spectra in the 280–650 Hz frequency range. Detection of the noise generated by the compromised membrane depended on the background noise.

Defects in Moving Parts

Mechanical defects in moving parts such as worn bearings produce high-frequency AE due to stress cracking, percussion or metal to metal contact. AE measurement has been used to detect mechanical defects before other symptoms, such as increased vibration and temperature or the appearance of metal particles in the lubricating oil, occur. Sundt (1979) attached acoustic sensors to the housing of a satisfactory bearing and a bearing with a hairline fracture. The good bearing produced normal background noise caused by the balls rolling in the races. The bad bearing produced high-amplitude spikes in the frequency spectrum which corresponded to the balls rolling over the crack. Li and Li (1995) used pattern recognition techniques to differentiate between good and damaged bearings.

Accident Detection within the Plant

Oikawa, Tomizawa, and Degawa (1997) developed an intelligent visual and acoustic monitoring system to detect accidents in a pant, which would reduce the need for maintenance operators to patrol the plant. Microphones were used to detect any unusual noises coming from rotary machines (forced draft fans, gas mixing fans and induced draft fans). The frequency domain 0–32kHz was divided into ten-octave bands. If the sound level exceeded a preset threshold level for various plant conditions in any octave band, an alarm would sound.

SUMMARY

In the introduction an ideal sensor was described as being non-intrusive, reliable, low costing and capable of real-time measurement for a wide range of process conditions. For acoustic measurement, direct contact with the process is not always essential allowing measurements to be made with a minimum of intrusion and in some cases completely non-intrusively. Acoustic measurement is also a real-time process and is therefore ideal for use in a control system.

Within this review examples have been given where the AE from processes have been investigated. Bubble sound in common gas–liquid dispersions has been investigated for characterising and monitoring bubble size, hold-up and mass transfer between phases. Solid mixing and transport processes also produce AE, which depend on process parameters such as particle size and mixing state. Certain chemical reactions have been shown to produce AE and these emissions have been used to classify and follow reactions.

The suitability of passive AE monitoring, for use in large plant-scale systems, depends on many factors. As already demonstrated there are many possible sources of AE within processes, including background sound from machinery such as pumps. In order to effectively analyse acoustic information, it is first necessary to separate the various sound sources. The AE described in studies discussed in this review tend to use volts or arbitrary units without quoting any calibration values so making comparisons based on their relative magnitudes impossible. In many cases rather than the magnitude of the AE involved, it is the frequency of the emissions that allow the different sound sources to be identified from the background noise. In Table 2, the various AE sources in processes discussed in the review and the frequency ranges at which they occur are summarised. The effects of scale on AE must also be investigated further. Increased size of equipment would generally result in the reduction of equipment vibration frequencies, vessel resonance frequencies and reduce wall effects such as echoes. This could aid in identifying and differentiating the various acoustic sources involved in the process. However, increased scale may also result in measurements being made further away from the acoustic source perhaps making detection more difficult.

Table 2: Summary of the various sources of sound in processes

Cause	Frequency of measured AE
Bubble formation	$\text{Frequency} \approx \dfrac{6.52}{\text{bubble diameter}}$ (Hz) (air bubbles in water)
Gas–liquid flow hydrodynamics	Measured up to 200Hz
Foam rupture	100–500kHz
Particle impact	15–200kHz
Fluidisation hydrodynamics	Up to 200Hz
Chemical reactions	50kHz upwards
Turbulence	Order of 10–100Hz

The economics of passive acoustic measurement are difficult to assess in general terms. However, the cost of acoustic sensors varies depending on their construction and calibration but it is similar to other measurement probes such as those used for dissolved oxygen. Extra costs might be imposed depending on the rate of signal acquisition and processing required. Passive acoustic measurement does not involve the generation of acoustic waves and therefore the equipment cost would be less than that of active acoustics.

The disadvantage in developing passive acoustic monitoring techniques is that they can only be applied to processes where AE are evolving and where the causes and sources of the emissions are understood. Knowledge of the acoustical properties of media and equipment involved in the process and how their properties change with process conditions is required. The difficulty in applying what in essence is a simple technique is in the interpretation of the signals, which are often very complicated, being made up of emissions from many acoustic sources. Due to the complex nature of the signals research is moving away from the fundamental understanding of the physics behind the AE towards using the AE as 'fingerprints' describing a specific set of conditions without necessarily knowing or understanding their cause. Principal component analysis allows the data to be condensed and neural network and multivariate models can be used to describe how the AE parameters vary with process conditions. The disadvantage of this approach is that the resulting models are generally case specific and must be developed for each application.

The examples discussed in this review demonstrate that there are many chemical engineering processes that cause AE. Despite any disadvantages, passive acoustic measurement is potentially a powerful process-monitoring tool, which could allow insight into the physicochemical state of the process where other techniques such as photography/video are inapplicable or require off-line, time-consuming analysis.

REFERENCES

1.   Asher, R. C. (1997). Ultrasonic sensors. Bristol: Institute of Physics Publishing.

2.   Belchamber, R. M., Betteridge, D., Collins, M. P., Lilley, T., Marczewski, C. Z., & Wade, A. P. (1986). Quantitative study of acoustic emission from a model chemical process. Analytical Chemistry, 58, 1873}1877.

3.   Belchamber, R. M., & Collins, M. P. (1989). A method of determining physical properties. EP0309155.

4.   Betteridge, D., Joshlin, M. T., & Lilley, T. (1981). Acoustic emissions from chemical reactions. Analytical Chemistry, 53, 1064}1073.

5.   Bouchard, J. G., Payne, P. A., & Szyszko, S. (1994). Non-invasive measurement of process states using acoustic emission techniques coupled with advanced signal processing. Transactions of the Institution of Chemical Engineers, 72A, 20}25.

6. Boyd, J. W. R., & Varley, J. (1997). Measurement of sound in bioreactors to estimate bubble size distributions. Fourth international conference on bioreactor and bioprocess yuid dynamics (pp. 379}394).

7. Boyd, J. W. R., & Varley, J. (1998). Sound measurement as a means of gas bubble sizing in aerated agitated tanks. A.I.Ch.E. Journal, 44(8), 1731}1739.

8. Brennen, C. E. (1995). Cavitation and bubble dynamics. Oxford: Oxford University Press.

9. Brereton, C. M. H., & Grace, J. R. (1992). The transition to turbulent #uidisation. Transactions of the Institution of Chemical Engineers, 70A, 246}251.

10. Brusewitz, G. H., & Venable, P. B. (1987). Technical notes: Sound level measurements of #owing grain. Transactions of the ASAE, 30, 863}864.

11. Cao, Z., Wang, B.-F., Wang, K.-M., Lin, H.-G., & Yu, R.-Q. (1998). Chemical acoustic emissions from gas evolution processes recorded by a piezoelectric transducer. Sensors and Actuators B, 50, 27}37.

12. Costigan, G., & Whalley, P. B. (1997). Measurements of the speed of sound in air}water #ows. Chemical Engineering Journal, 66, 131}135.

13. Crowther, T. G., Wade, A. P., Wentzell, P. D., & Gopal, R. (1991). Characterization of acoustic emission from an electrolysis cell. Analytica Chimica Acta, 254, 223}234.

14. De More, L. S., Pa!ord, W. F., & Tatterson, G. B. (1988). Cavity sound resonance and mass transfer in aerated agitated tanks. A.I.Ch.E. Journal, 34(11), 1922}1926.

15. Dhodapkar, S. V., & Klinzing, G. E. (1993). Pressure #uctuations in pneumatic conveying systems. Powder Technology, 74, 179}195.

16. Drahos, J., & Cermak, J. (1989). Diagnostics of gas}liquid #ow patterns in chemical engineering systems. Chemical Engineering and Processes, 26, 147}164.

17. Drahos, J., Cermak, J., Selucky, K., & Ebner, L. (1987). Characterization of hydrodynamic regimes in horizontal two-phase #ow part II: Analysis of wall pressure #uctuations. Chemical Engineering Processes, 22, 45}52.

18. Drahos, J., Zahradnik, J., Puncochar, M., Fialova, M., & Bradka, F. (1991). E!ect of operating conditions on the characteristics of pressure #uctuations in a bubble column. Chemical Engineering Processes, 26, 147}164.

19. Drouillard, T. F. (1988). Introduction to acoustic emission. Materials Evaluation, 46, 175}180.

20. Elsey, J., Barton, G. W., Jungk, S., Francis, G., Sellahewa, J., & Chessari, C. (1998). Acoustics based on-line quality estimation. Computers in Chemical Engineering, 22, 5925}5928.

21. Folkestad, T., & Mylvaganam, K. S. (1990). Acoustic measurements detect sand in North Sea #ow lines. Oil and Gas, 33}39.

22. Fowler, T. J. (1992). Chemical industry applications of acoustic emission. Materials Evaluation, 875}882.

23. Glasgow, L. A., Erickson, L. E., Lee, C. H., & Patel, S. A. (1984). Wall pressure #uctuations and bubble size distributions at several positions in an airlift fermentor. Chemical Engineering Communications, 29, 311}336.

24. Glasgow, L. A., Hua, J., Yiin, T. Y., & Erickson, L. E. (1992). Acoustic studies of interfacial e!ects in airlift reactors. Chemical Engineering Communications, 113, 155}181.

25. Hakanen, A., & Laine, E. (1995). Acoustic characterization of microcrystalline cellulose powder during and after its compression. Drug Development and Industrial Pharmacy, 21(13), 1573}1582.

26. Hakanen, A., Laine, E., Jalonen, H., Linsaari, K., & Jokinen, J. (1993). Acoustic emission during powder compaction and its frequency spectral analysis. Drug Development and Industrial Pharmacy, 19(19), 2539}2560.

27. Harrenstein, A., & Brusewitz, G. H. (1987). Sound level measurements of #owing wheat. Transactions of the ASAE, 29, 1114}1117.

28. Hou, R., Hunt, A., & Williams, R. A. (1998). Acoustic monitoring of hydrocyclone performance. Minerals Engineering, 11(11), 1047}1059.

29. Hou, R., Hunt, A., & Williams, R. A. (1999). Acoustic monitoring of pipeline #ows: Particulate slurries. Powder Technology, 106, 30}36.

30. Hsi, R., Tay, M., Bukur, D., Tatterson, G. B., & Morrison, G. (1985). Sound spectra of gas dispersion in an agitated tank. Chemical Engineering Journal, 31, 153}161.

31. Hunaida, O., & Chu, W. T. (1999). Acoustical characteristics of leak signals in plastic water distribution pipes. Applied Acoustics, 58, 235}254.

32. Johnsson, F., Zijerveld, R. C., Schouten, J. C., van den Bleek, C. M., & Leckner, B. (2000). Characterization of #uidization regimes by time-series analysis of pressure #uctuations. International Journal of Multiphase Flow, 26, 663}715.

33. Kinsler, L. E. (1982). Fundamentals of acoustics (3rd ed.). New York, Chichester: Wiley.

34. Kolb, W. B., Melli, T. R., de Santos, J. M., & Scriven, L. E. (1990). Cocurrent down#ow in packed beds. Flow regimes and their acoustic signatures. Industrial and Engineering Chemistry Research, 29, 2380}2389.

35. Krieg, D. A., Helwick, J. A., Dillon, P. O., & McCready, M. J. (1995). Origin of disturbances in cocurrent gas}liquid packed bed #ows. A.I.Ch.E. Journal, 41(7), 1653}1666.

36. Kupferberg, A., & Jameson, G. J. (1970). Pressure yuctuations in a bubbling system with a special reference to sieve plates, 48, 140}150.

37. Laine, J. M., Glucina, K., Chamant, M., & Simonie, P. (1998). Acoustic sensor: A novel technique for low pressure membrane integrity monitoring. Desalination, 119, 73}77.

38. Leach, M. F., & Rubin, G. A. (1978). Size analysis of particles of irregular shape from their acoustic emissions. Powder Technology, 21, 263}267.

39. Leach, M. F., Rubin, G. A., & Williams, J. C. (1977). Particle size determination from acoustic emissions. Powder Technology, 16, 153}158.

40. Leach, M. F., Rubin, G. A., & Williams, J. C. (1978a). Particle size distribution characterization from acoustic emissions. Powder Technology, 19, 157}167.

41. Leach, M. F., Rubin, G. A., & Williams, J. C. (1978b). Analysis of a Gaussian size distribution of rigid particles from their acoustic emission. Powder Technology, 19, 189}195.

42. Leighton, T. G. (1994). The acoustic bubble. New York: Academic Press.

43. Leighton, T. G., Fagan, K. J., & Field, J. E. (1991). Acoustic and photographic studies of injected bubbles. European Journal of Physics, 12, 77}85.

44. Li, C. J., & Li, S. Y. (1995). Acoustic emission analysis for bearing condition monitoring. Wear, 185, 67}74.

45. Lubetkin, S. D. (1989). Measurement of bubble nucleation rates by an acoustic method. Journal of Applied Electrochemistry, 19(5), 668}675.

46. Manasseh, R. (1996). Acoustic sizing of bubbles at moderate to high bubbling rates. Fourth world conference on experimental heat transfer yuid mechanics and thermodynamics 2}6 June.

47. Manasseh, R., Lafontaine, R. F., Davy, J., Sheperd, I., & Zhu, Y.-G. (2000). Passive acoustic bubble sizing in sparged systems. Experiments in Fluids, submitted for publication.

48. McClements, D. J. (1997). Ultrasonic characterisation of foods and drinks: Principles, methods, and applications. Critical Reviews in Food Science and Nutrition, 37(1), 1}46.

49. Minnaert, M. (1933). On musical air-bubbles and the sounds of running water. Philosophical Magazine, 16, 235}248.

50. Newland, D. E. (1994). An introduction to random vibrations, spectral and wavelet analysis (3rd ed.). UK: Longman Scienti"c and Technical.

51. Oikawa, T., Tomizawa, M., & Degawa, S. (1997). New monitoring system for thermal power plants using digital image processing and sound analysis. Control Engineering Practice, 5(1), 75}78.

52. Pandit, A. B., Varley, J., Thorpe, R. B., & Davidson, J. F. (1992). Measurement of bubble size distribution: An acoustic technique. Chemical Engineering Science, 47(5), 1079}1089.

53. Park, Y., Lamont Tyler, A., & de Nevers, N. (1977). The chamber ori"ce interaction in the formation of bubbles. Chemical Engineering Science, 32, 907}916.

54. Parkinson, G. (1991). Holistic maintenance. Chemical Engineering, 30}35.

55. Roy, R., Davidson, J. F., & Tuponogov, V. G. (1990). The velocity of sound in #uidised beds. Chemical Engineering Science, 45, 3233}3245.

56. Rzesotarska, J., Rejmund, F., & Ranachowski, P. (1998). Acoustic emission measurement of foam evolution in $H_2 O\}C_2 H_5 OH\}$air systems with content of detergent triton X-100. Ultrasonics, 36, 953}958.

57. Sawada, T., Gohshi, Y., Chikako, A., & Furuya, K. (1985a). Acoustic emissions arising from the gelation of sodium carbonate and calcium chloride. Analytical Chemistry, 57, 366}367.

58. Sawada, T., Gohshi, Y., Chikako, A., & Furuya, K. (1985b). Acoustic emission from phase transition of some chemicals. Analytical Chemistry, 57, 1743}7145.

59. Shepard, D. D., & Smith, K. R. (1997). A new ultrasonic measurement system for the cure monitoring of thermosetting resins and composites. Journal of Thermal Analysis, 49, 95}100.

60. Strasberg, M. (1956). Gas bubbles as sources of sound in liquids. Journal of the Acoustical Society of America, 28(1), 20}26.

61. Sundt, P. T. (1979). Monitoring acoustic emission to detect mechanical defects. InTech, 43}44.

62. Sutter, T. A., Morrison, G. L., & Tatterson, G. B. (1987). Sound spectra in an aerated agitated tank. A.I.Ch.E. Journal, 33(4), 668}671.

63. Tily, P. J., Porada, S., Scruby, C. B., & Lidington, S. (1987). Monitoring of mixing processes using acoustic emission, #uid mixing III, Bradford, September 8}10, Symposium Series no. 108 (pp. 75}94).

64. Turner, J. D., & Pretlove, A. J. (1991). Acoustics for engineers. Basingstoke: Macmillan.

65. Usry, W. R., Morrison, G. L., & Tatterson, G. B. (1987). On the interrelationship between mass transfer and sound spectra in an aerated agitated tank. Chemical Engineering Science, 42(7), 1856}1859.

66. Vahaviolos, S. J., Pollock, A., & Lew, N. (1991). Pinpoint structural defects with acoustic emissions. Chemical Engineering Progress, 87(1), 60}67.

67. van der Schaaf, J., Schouten, J. C., & van den Bleek, C. M. (1998). Origin, propagation and attenuation of pressure waves in gas} solid #uidized beds. Powder Technology, 95, 220}233.

68. Van Ooijen, J. A. C., van Tooren, E., & Reedijk, J. (1978). Acoustic emission during the preparation of dichloro(pyrazine)zinc(II). Journal of the American Chemical Society, 100(17), 5569}5570.

69. Watkins, R. R. D., Haywood, R. B. C. G., & Scruby, C. B. (1989). Acoustic monitoring of plant operation. GB Patent, 2211938.

70. Wentzell, P. D., Lee, O., & Wade, A. P. (1991). Comparison of pattern recognition descriptors for chemical acoustic emission analysis. Journal of Chemometrics, 5, 389}403.

71. Wentzell, P. D., & Wade, A. P. (1989). Chemical acoustic emission analysis in the frequency domain. Analytical Chemistry, 61, 2638}2642.

72. Zukowski, W. (1999). Acoustic e!ects during the combustion of gaseous fuels in a bubbling #uidized bed. Combustion and Flame, 117, 629}635.

Investigation of Inlet Gas Streams Effect on the Modified Claus Reaction Furnace

Reza Rezazadeh[1] and Sima Rezvantalab[2]

[1]Shahid Hasheminejad Gas Processing Company (S.G.P.C), Khangiran, Iran

[2]Department of Chemical Engineering, Urmia University of Technology, Urmia, Iran

ABSTRACT

The objective of this paper is to model the main reactions that take place in the Claus reactor furnace and compare it with actual data and simulated process. Since the most important point is the selection of suitable reaction conditions to increase the reactor performance, the model is formulated to predict the performance of the Claus plant. To substantiate the theoretical model, we used actual process condition and feed composition in Shahid Hasheminejad Gas Refinery. Model equations have been solved by using MATLAB program. Results from MATLAB are compared with those from *SULSIM* $^\rightarrow$ simulator and with actual plant data. The AAD (Average Absolute Deviation) of modeling results with actual data is 2.07% and AAD of simulation results with real data is 4.77%. Error values are very little and show accuracy and precision of modeling and simulation. The predicting curve for different parameters of the reactor furnace according to variable conditions and specifications are given.

INTRODUCTION

Hydrogen sulfide (H_2S) is produced from sulfur compounds in fossil fuels such as natural gas or oil. Sour gases (H_2S and CO_2), are removed from the natural gas or refinery gas by means of one of the gas treating processes. Due to global environmental rules, refineries have to recover sulfur from nature. H_2S containing acid gas stream is flared, incinerated, or fed to a sulfur recovery unit. The Claus process is commonly used to reduce the emission of sulfur compounds into the atmosphere. Recently recovery of sulfur is done by means of the modified Claus tail gas clean-up processes. In these processes, H_2S over a catalyst converts to elemental sulfur where the reaction takes place in a high temperature furnace. The recovery process is the reaction between H_2S and air to form sufur and water. Following reaction is the main reaction in the recovery process:

$$H_2S + \frac{1}{2}O_2 \rightarrow \frac{1}{x}S_x + H_2O + \text{heat}\left(\cong 626000 \frac{KJ}{Kgmole} \right) \tag{1}$$

In the original Claus process, control of this reaction was difficult and sulfur recovery efficiencies were low. In order to overcome these difficulties and also increase the efficiency of the process, several modifications of the Claus process have been developed. In modified process, free flame total oxidation of 1/3 of the H_2S to SO_2 followed by a reaction over the catalyst of SO_2 with the remaining 2/3 H_2S. According to Mohamed Sassi and Ashwani K. Gupta modified Claus process for a Sulfur Recovery Plant consists of several stages [1]:

Combustion (In the Reactor Furnace)

$$H_2S + 3/2O_2 \xrightarrow{\text{High T}} SO_2 + H_2O$$
$$+ \text{heat}\left(\cong 518000 \ KJ/Kgmole \right) \tag{2}$$

$$2H_2S + SO_2 \xleftrightarrow{\text{High T}} 3/2\,S_2 + 2H_2O$$

$$\Delta H_R \cong +47000 \ \text{KJ/Kgmole} \tag{3}$$

Redox (Catalytic Converter)

These are simplified reactions which actually take place in a Claus unit. There are various species of gaseous sulfur S_2, S_3, S_4, S_5, S_6, S_7, and S_8. Equilibrium concentrations of these sulfur compounds are not known in the entire of process. Additionally, gas stream contains water saturated with 15 - 80 mol% H_2S, 0.5 - 1.5 mol% hydrocarbons, and CO_2 which can result in carbonyl sulfide (COS) carbon disulfide (CS_2), carbon monoxide (CO), and hydrogen [2].

$$2H_2S + SO_2 \xleftrightarrow{\text{Catalytic, Low T}} \frac{3}{X}S_x + 2H_2O + \text{heat}$$

$$\left(\cong 108 \ \text{KJ/Kgmole} \right) \tag{4}$$

$$3H_2S + 3/2\,O_2 \xrightarrow{\text{Overall}} \frac{3}{X}S_x + 3H_2O + \text{heat}$$

$$\left(\cong 626000 \ \text{KJ/Kgmole} \right) \tag{5}$$

Most Claus plants operate in the multistep process "straight-through" mode as shown in Figure 1. The combustion is carried out in a reducing atmosphere with only enough air 1) to oxidize one-third of the H_2S to SO_2, 2) to burn hydrocarbons and mercaptans, and 3) for many refinery Claus units, to oxidize ammonia and cyanides. The process includes the following operations:

- Combustion: burn hydrocarbons and other combustibles and oxidize one-third of the H_2S to provide necessary SO_2 to react with remainder H_2S for producing S_2 in the furnace.
- Waste Heat Recovery: Cool combustion products.
- Sulfur Condensing: Cool outlet streams from waste heat recovery unit and from catalytic converters.
- Reheating: Reheat process stream, after sulfur condensation and separation, to a temperature high enough to remain sufficiently above the sulfur dew point.

In order to investigate different aspects of the modified Claus process, a number of studies have been performed on main burner and sulfur recovery in this process. Monnery et al. modeled the modified

Claus process [3]. Kelly Anne Humboldt has studied mathematical modeling of reactions in the process [4]. Recently, S. Asadi et al. used TSWEET simulator to optimize the recovery of sulfur [5].

At first approach, we have used a mathematical model for the key reactions that take place in the reactor furnace. In the second approach, we have simulated the process with a commercial simulator. Finally using the model and simulation, we have compared obtained results and proposed some improvements on the base case.

KINETIC STUDIES

Claus process has been investigated via different aspects, experimental and theoretical perspectives. Paskal et al. gives a summary of the main reactions thought to occur within the Claus furnace [6, 7]. Clark et al. discussed the mechanisms behind the formation of key sulfur containing species found within the furnace, and in a subsequent study outline primary reaction pathways for the principal components in the furnace [8, 9]. While there have been numerous attempts to model the Claus process based on simplified kinetic expressions, the complexity of the chemistry and the number of involved reactions has precluded the accurate prediction of outlet compositions. As it mentioned before, gas stream contains different compounds such as H_2S, CO_2 and hydrocarbons. Most important compound is H_2S and several groups have studied it's decomposition under different condition. As result, it suggested that there are numerous reactions on the catalytic decomposition of H_2S in the clause process. According to the studies, gaseous H_2S exists in chemical equilibrium with elemental hydrogen and sulfur by the following equation:

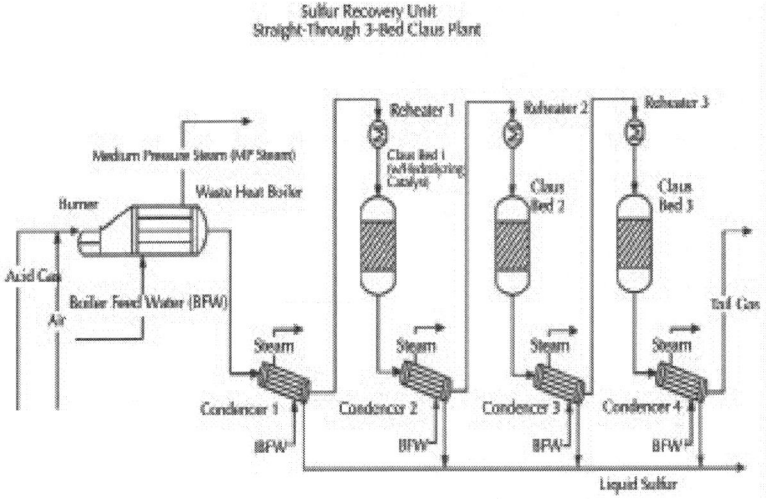

Figure 1: Three-stage straight-through sulfur plant.

$$H_2S \leftrightarrow H_2 + \frac{1}{x}S_x$$

(6)

Oxidation includes two staged reaction, first oxidation of H_2S and followed by reaction between H_2S and SO_2 that limiting stage of the Claus reaction is the second part [10].

$$H_2S + \frac{3}{2}O_2 \leftrightarrow H_2O + SO_2$$

(7)

$$2H_2S + SO_2 \leftrightarrow 2H_2O + S_2$$

(8)

During the reaction in the furnace and according to the existence of ammonia in the gas stream, oxidation of NH_3 will take place. Recently, Clark has mentioned that under 1100°C ammonia oxidation is negligible. Additionally, he noted in competitive oxidation, first of all H_2S, and then methane, finally NH_3 react. On the other hand, Goar et al. found rate of hydrocarbon combustion is more than ammonia and NH_3 is more than H_2S [11]. Formation of COS and CS_2 in the Claus reaction furnace are also very important in the modeling. Field studies have revealed that concentrations of COS and CS_2 at the exit of the reaction furnace/ waste heat boiler typically lie between 100 ppm and

2 mol% [12]. However, these seemingly small concentrations in the furnace product stream can represent nearly half the sulfur emissions from a tail gas clean-up unit [13]. It is possible to hydrolyze COS and CS_2 back into H_2S in the Claus catalytic converters according to the following stoichiometry:

$$COS + H_2O \leftrightarrow CO_2 + H_2S \tag{9}$$

$$CS_2 + 2H_2O \leftrightarrow CO_2 + 2H_2S \tag{10}$$

As it mentioned, there many reactions which may take place in the furnace according to the conditions such as temperature and pressure. Full list of reactions that occur in the furnace is not obvious and for the known reactions, reaction rate expressions are not available. In current work we assumed that gas stream consists of CH_4, CO_2, H_2S, H_2O, O_2, N_2, CO, CS_2, COS, S_2, SO_2, H_2. Regarding gas stream composition, important reactions which take place in furnace and we use in the modeling are listed below.

$$H_2S + 3/2 O_2 \rightarrow SO_2 + H_2O \tag{11}$$

$$2H_2S + SO_2 \leftrightarrow 3/2 S_2 + 2H_2O \tag{12}$$

$$CH_4 + 2O_2 \rightarrow CO_2 + 2H_2O \tag{13}$$

$$CO_2 \rightarrow CO + 1/2 O_2 \tag{14}$$

$$H_2S \rightarrow 1/2 S_2 + H_2 \tag{15}$$

$$CH_4 + 2S_2 \rightarrow CS_2 + 2H_2S \tag{16}$$

$$CO + 1/2 S_2 \rightarrow COS \tag{17}$$

MATHEMATICAL MODEL

The basic structure of the model consist of the equations of mole and energy conservative rule the furnace, which are related to each other and are function of molar conversion of H_2S in equilibrium reaction and temperature. In order to model the reactor, a steady-state simulation has been used for mole and energy balance.

Sames and Paskal presented empirical correlations to predict the fraction of CO, H_2, COS, CS_2 and sulfur (as S) in the effluent of the Claus furnace. The correlations are obtained from more than 300 tests on 100 different sulfur trains; with different flow configurations processing acid gas feed streams [12]. These empirical correlations are presented in Appendix. We use these equations to model the furnace and mole balance. In this work, furnace pressure is 130 kPa (absolute) and pressure drop (ΔP) is 10 kPa. Using empirical equations and applying in the mole balance for the compounds, we get the mole balance equations, for each compound.

For gas components in the outlet gas stream:

H_2S:

$$f_{i(H_2S)} + 2R(CS_2) \times \left(f_{i(CH_4)} \right) - f_{i(H_2S)} - \frac{1}{3} f_{i(H_2S)}$$

$$-R(H_2) \times f_{i(H_2S)} - X \tag{18}$$

H_2O:

$$f_{i(H_2O)} + 2\left(1 - R(CS_2)\right) \times f_{i(CH_4)} + \frac{1}{3} f_{i(H_2S)} + X \tag{19}$$

CO_2:

$$f_{i(CO_2)} + f_{i(CH_4)} \times \left(1 - R(CS_2)\right)$$

$$-\left(R(COS) + R(CO) \right) \times f_{i(CH_4)} + f_{i(CO_2)} \tag{20}$$

CO:

$$R(CO) \times \left(f_{i(CH_4)} + f_{i(CO_2)} \right) \tag{21}$$

SO_2:

$$\frac{1}{3} f_{i(H_2S)} - \frac{1}{2} X \tag{22}$$

CH_4:

$$0 \tag{23}$$

O_2:

$$\left(R(\text{COS})\right)+R(\text{CO})\times\left(f_{i(\text{CH}_4)}+f_{i(\text{CO}_2}\right. \tag{24}$$

N_2:

$$\frac{79}{21}f_{i(\text{H}_2\text{S})}+2f_{i(\text{CH}_4)}\left(1-R(\text{CS}_2)\right)\times\frac{1}{2} \tag{25}$$

S_2:

$$\frac{3}{4}XR(\text{COS})\times\left(f_{i(\text{CH}_4)}+f_{i(\text{CO}_2)}\right)+\frac{1}{2}f_{i(\text{H}_2\text{S})}$$

$$-2R(\text{CS}_2)\times f_{i(\text{CH}_4)}-\frac{1}{2}R(\text{H}_2) \tag{26}$$

H_2:

$$R(H_2)\times f_{i(\text{H}_2\text{S})} \tag{27}$$

CS_2:

$$R(\text{CS}_2)\times f_{i(\text{CH}_4)} \tag{28}$$

COS:

$$R(\text{COS})\times\left(f_{i(\text{CH}_4)}+f_{i(\text{CO}_2)}\right) \tag{29}$$

Total moles:

$$f_{i(\text{CH}_4)}\left(3-2R(\text{CS}_2)+\frac{1}{2}R(\text{CO})\right)$$

$$+f_{i(\text{CO}_2)}\left(1-R(\text{CS}_2)+\frac{1}{2}R(\text{CO})\right)$$

$$+f_{i(\text{H}_2\text{S})}\left(\frac{4}{3}+\frac{1}{2}R(\text{H}_2)\right)+f_{i(\text{N}_2)}+f_{i(\text{H}_2\text{O})}+\frac{1}{4}X \tag{30}$$

X: Flow rate of H_2S conversion at the equilibrium Claus reaction at equilibrium:

$$K_p = \frac{[H_2O]^2 [S_2]^{3/2}}{[H_2S]^2 [SO_2]} \left[\frac{\pi}{total \; mols} \right]^{3/2-1}$$ (31)

By interpolation in the K_p curve vs. furnace temperature (GPSA), we calculate:

$$K_p = 0.6021T - 420.49T \left(°C \right), K_p \left(Kpa^{0.5} \right)$$ (32)

Replacing molar flows in equation (18) through 30 in equation (31) resulted into a new equation, function of temperature, in the following form (see below):

Where the A, B, C, D are function of feed composition and would be calculated numerically.

$$T(°K) = \left(\left(\frac{(A+X)^2 (B+0.75X)^{1.5} \pi^{0.5}}{(C-X)^2 \left(1/3 f_{t(H_2S)} - 0.5X \right)(D+0.25)^{0.5}} + 420.49 \right) \Big/ 0.6021 \right) + 273$$ (33)

In order to calculate the reactions enthalpy, we used equation 34 and data provided in table 1. Table 1 represents standard enthalpy of formation and heat capacity parameters for each component.

$$SUM \left(\Delta H_{rxn} \right) = \sum_{1}^{7} \Delta H_{rxn} \times f_{rxn(i)} \times Conversion_i$$ (34)

ΔH_{rxn} is standard enthalpy of reactions in 25°C, $f_{rxn(i)}$ is molar flow rate of limiting compound in each reaction, $Conversion_i$ is conversion value of limiting compound in each reaction.

Replacing all mentioned equations resulted into a new equation (35) in which conversion of H_2S is function of temperature of reaction furnace (table 2).

$$X = \frac{A_0 + A_4T + B_4T^2 + C_4T^3 + D_4T^4}{B_0 + A_5T + B_5T^2 + C_5T^3 + D_5T^4}$$ (35)

A_0, A_4, A_5, B_0, B_4, B_5, C_4, C_5, D_4, D_5 are constant parameters that would calculated numerically according to the inlet gas stream condition.

In order to establish a reference point, calculations are carried out for a "base case" and the operating conditions used are given in Table 3. Our base case is Shahid Hasheminejad Gas Refinery. Shahid Hasheminejad (Khangiran) gas refinery is in Sarakhs, Khorasan province.

Operating conditions is composition of inlet gas stream to the Claus process in the refinery. Next step, application of the feed gas condition in the equations resulted into the parameters of mole and energy balance equations. Calculated parameters are presented in table 4.

PROCESS SIMULATION

In order to compare the implications of the reaction furnace model on design, overall sulfur recovery and emissions, this modified Claus plant was simulated using $SULSIM^{\rightarrow}$.

Table 1: Compound heat capacity parameters and H_{298}^0 kJ/kg mole

	a_i	b_i	c_i	d_i	$H_{i,298}^0$ at 25°C kJ/kgmole
CH_4	34.31	0.05469	3.66×10^{-6}	-11×10^{-9}	881
CO_2	19.8	0.0734	-5.6×10^{-5}	1.72×10^{-8}	912
H_2S	31.94	1.44×10^{-3}	$2.43 \times 1T^8$	-1.18×10^{-8}	846
H_2O	32.2	1.92×10^{-3}	1.06×10^{-5}	3.6×10^{-9}	839
0_2	29.1	11.58×10^{-3}	$-6.0759 \times 1e$	13.11×10^{-16}	733
N_2	21.15	13.6×10^4	2.6×10^{-5}	-1.17×10^{-8}	726
SO_2	23.85	67×10^{-3}	-4.9×10^{-5}	$1.33 \times 1 \ 10^{-8}$	986
S_2	32.47	0.0067	0	0	805
CO	30.87	-12.9×10^{-3}	2.79×10^{-5}	-1.27×10^{-8}	729
$CS:$	27.44	81.3×10^{-3}	-7.67×10^{-5}	2.87×10^{-8}	1127

COS	23.57	79.8x10^{-3}	-7.02x10^{-5}	2.45x10^{-8}	1025
H$_2$	27.14	9.27x10^{-3}	-1.38x10^{-5}	7.65x10^{-9}	719

Table 2: limiting specie and calculated H$_{rxn}$ [4]

Limiting specie	Reactions	ΔH$_{rxn}$ (kj/kgmole)
CO$_2$	$CO_2 \rightarrow CO + 1/2O_2$	280000
H$_2$S	$H_2S + 3/2O_2 \rightarrow SO_2 + H_2O$	-520000
CH$_4$	$CH_4 + 2O_2 \rightarrow CO_2 + 2H_2O$	-803000
H$_2$S	$2H_2S + SO_2 \leftrightarrow 3/2S_2 + 2H_2O$	46500
H$_2$S	$H_2S \rightarrow 1/2S_2 + H_2$	81000
CH$_4$	$CH_4 + 2S_2 \rightarrow CS_2 + 2H_2S$	-296000
CO	$CO + 1/2S_2 \rightarrow COS$	-142000

Table 3: Refinery Claus process inlet gas condition

Feed Temperature	325 K (52°C)
Inlet air temperature (average in summer and winter)	350 K (77°C)
Furnace pressure	130 kPa A
Feed molar rate	2682.9 Kgmole/h
CH$_4$ mole fraction	0.0105
CO$_2$ mole fraction	0.563
H$_2$S mole fraction	0.336
H$_2$O mole fraction	0.0905

Table 4: Calculated parameters for case study

i	A_i	B_i	C_i	D_i
-	560.6297	-6.5079	589.4866	3748.3
0	2.0757×10^8	4.3965×10^4		
4	-1.1318×10^5	-68.2566	0.0115	-2.3178×10^{-7}
5	12.6875	0.014	3.7×10^{-6}	3.875×10^{-10}

SULSIM$^{\rightarrow}$ is program for Sulfur Plant Simulation and represents simulation package for SRU and TGTU design. This software has widely accepted thermodynamic data and propriety thermodynamic properties for all of the gas components and sulfur species found in sulfur recovery processes. Figure 2 shows the flow diagram of the simulated Claus unit of Shahid Hasheminejad Gas Refinery.

RESULTS AND DISCUSSION

As described previously, we implemented real data from gas refinery in the model. In this section to verify the model, we compare the output result from reaction furnace, model values and simulation results. Table 5 listed three set of results.

It is obvious that furnace temperature obtained using model (1098 K) is lower than actual temperature of reaction furnace (1113 K), as the 15°C temperature difference is negligible and it results into 1.35% error.

Simulated results shows that predicted temperature using *SULSIM*$^{\rightarrow}$ software is 1121 K and higher than Claus furnace temperature. Error occurred using software is lower (0.72%). In this case simulation is more reliable. Additionally, sulfur conversion obtained from model results (56.635%) is in good agreement with conversion of sulfur in Claus plant (54%) in gas refinery. On the other hand, results from simulation indicate that sulfur conversion is 60.28%. According to the fact that obtained conversions from model are attained in lower temperature in comparing with the plant temperature while sulfur product is higher, we can conclude that model is more efficient in sulfur conversion. It is obvious that presented model in these conditions, is effective even more than simulated process. As it is presented in the table 5, in the furnace effluent there is sulfur vapor. It's due to the high temperature in the furnace; this high temperature converts sulfur to S_2 vapor. According to table 5, inlet air ratio to acid gas feed in Claus plant is 0.86 and in our model this ratio is 0.83. As we used this obtained ratio in our simulation, simulation and modeling resulted in the same ratio. Therefore modeling and simulation errors are low and about 3.5%. Since sulfur production is high in comparing with actual plant; air consumption is low, CO_2 concentration difference is about 0.05 (mole %), it means error is 0.34%; we can conclude that in all cases, simulated and modeling are more efficient.

Predicted N_2 concentration using equilibrium model is 36.067 (mole %) whereas N_2 content in the plant outlet gas stream is 39 (mole %). Since In empirical equations of model, it is assumed air consumption is low, results are logic and difference between real state and model results is acceptable. Since air consumption is low, its acid gas capacity is more than actual plant and can predict better results. Simulation has the same manner in the prediction of N_2 concentration. H_2O concentration in the outlet stream from model is less than plant outlet water content. The model performance was not good in water case and error value is about 7.8%. While H_2O content in simulation is closer to the actual data and lower error has occurred. Simulation performance is better in this case.

O_2 component in Claus plant damages the equipments (catalyst exchanger) and must be minimized, in Claus plants O_2 content is zero. Predicted concentration is 0.014% and error value is acceptable. The model predicted CH_4 concentration would drop to zero in the outlet of furnace. Checking the actual outlet concentration it is obvious that there is no methane in outlet stream, therefore no error has occurred in predicting of methane concentration. Table 5 represents actual data for the concentration of CS_2 and COS are greater than equilibrium model, data obtained from simulation. As you know, production of CS_2 and COS in the Claus furnace reduces the efficiency of total plant and it is better to decrease these components. Therefore simulation and model outputs are more effective. According to Sames (1990), COS forms in WHB exchanger, therefore difference between predicted concentration and plant data verifies the formation reaction of COS [14]. This take place in W.H.B as following equation:

$$CO + 1/2\,S_2 \rightarrow COS \qquad\qquad (36)$$

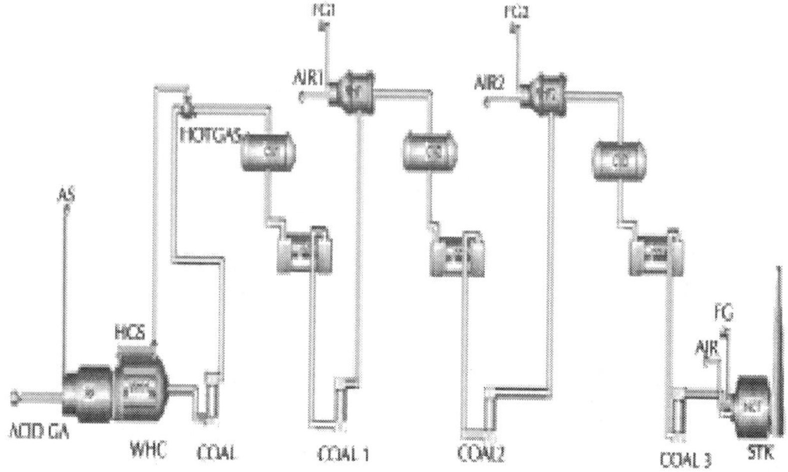

Figure 2: *SULSIM*$^\rightarrow$ simulation used for gas for Khangiran gas refinery (S.G.P.C).

Table 5: Comparison of plant value and results from model and simulation for S.G. P.C

Condition of Outlet Stream	Actual values for outlet concentration from W.H.B in Claus unit	Predicted Values using Model	Predicted Values using SULSIM® simulation
T (furnace temp)	1113 K (840°C)	1098 K (825°C)	1121 K (848°C)
F_{out} (kgmole/hr)	4789.7	4879.593	4882.196
X_{CH4}	0	0	0
X_{CO2} (mole %)	31	31.105%	30.698%
X_{H2S} (mole %)	4.9%	5.051%	4.952%
X_{H20} (mole %)	20.1%	18.519%	19.214%
X_{02} (mole %)	0%	0.014%	0

X_{N2} (mole %)	39%	36.067%	36.076%
X_{SO2}	2.8%	2.643%	2.301%
X_{s2} (mole %)	1%	5.139%	5.516%
X_{00} (mole %)	0.2%	0. 325%	0.737%
$X_{COS, CS2}$ (mole %)	0.9%	0.102%	0.081%
X_{H2} (mole %)	0.1%	1.035%	0.425%
Ratio (air/feed)	0.86	0.83	0.83
Sulfur conversion	54%	55.635%	60.28%

Predicted S_2 content in both methods is greater than plant data. There is a big difference between real and predicted values for S_2; it is due to the formation of liquid sulfur in WHB. According to table 5 and comparison between results and plant data, and also neglecting the error in CO and H_2 predicted concentrations, average error is about 3.5% and 5.36% for model and *SULSIM*⁻ simulation; also AAD (Average Absolute Deviation) in comparing actual data with modeling and simulation results are 2.07% and 4.92%, respectively. We can conclude that our model is more efficient and applicable for other Claus plants with different inlet composition.

H₂S Concentration Effects

Figure 3 shows reaction furnace predicted temperature vs. inlet H_2S content using model and simulation. Both simulation and model have similar trend. According to the figure, model predicts that 1% increase in H_2S content will result into 7.5°C increase in furnace temperature. In order to combust hydrocarbons and aromatics, furnace temperature must be 1050°C. According to the model, if inlet gas stream contains more than 26% H_2S in current plant, temperature would increase to higher than 1050°C (1323 K). Figures 4(a) and (b) illustrate the predictions of model and simulation for H_2S and sulfur conversion in furnace vs. increase in H_2S content in feed. As model predicts, for one percent of the mole fraction of H_2S in feed stream, sulfur conversion increases by 0.54% in reaction furnace and S_2 mole fraction in outlet gas stream increases 0.12%. Simulation has similar manner in this case. Obtained results both are matched. It should be noted that sulfur conversion in figure 4(b) shows model and simulation have similar trend. Therefore there is a negligible difference between the model and simulation prediction in 18% H₂S in figure 4(a). It is due to errors occurred in the simulation.

Figure 5 shows the effect of H₂S concentration in feed stream on the effluent H₂S content. As H₂S content increase in the feed, model predicts a trend for H₂S content in outlet which decreases and then with a lower slope increases and finally decreases.

The results indicate that in lower concentrations, furnace temperature is low and increase in the H₂S (to 30%) content would increase the slope of H₂S conversion line that results to decrease in H₂S content in the outlet stream of furnace.

In the higher concentrations, since furnace temperature is more than previous, H₂S cracking and conversion in Claus reaction increases and unconverted hydrogen sulfide in the outlet decreases. As can be seen in the figure, the reduction of H₂S in the effluent is agreement with increase in H₂S content in acid feed gas. For this case, model prediction is more reliable than simulation.

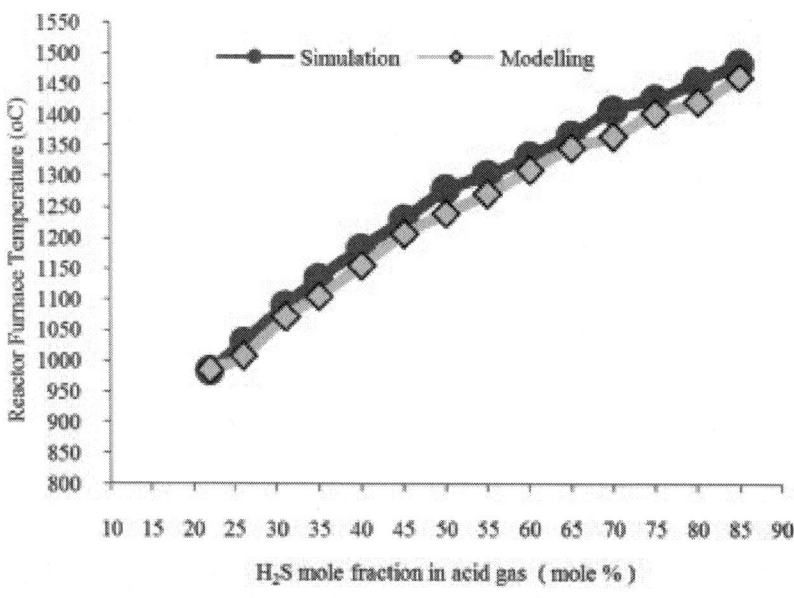

Figure 3: Simulation and model estimation for Modified Claus plant reaction furnace temperature vs. H₂S content.

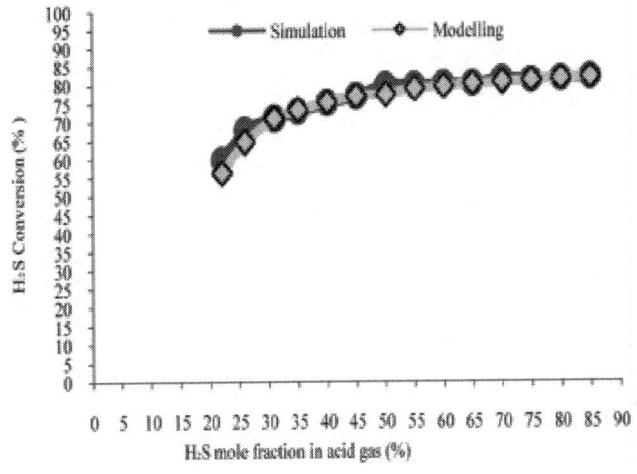

(a)

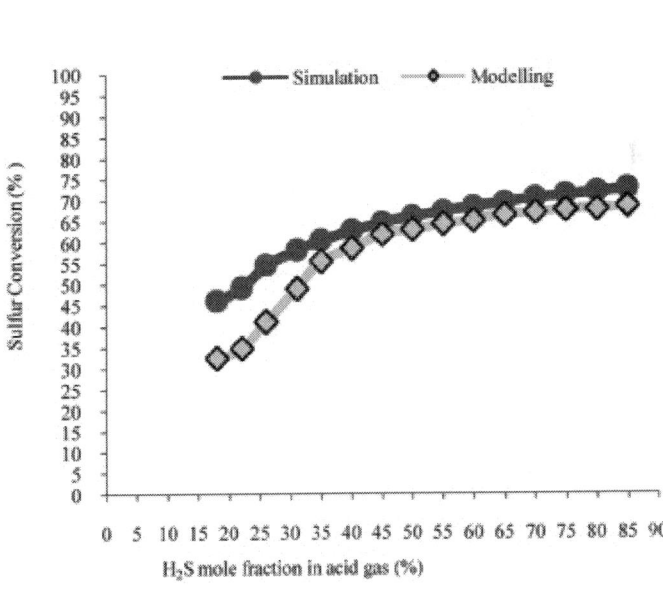

(b)

Figure 4: Comparison of predicted. (a) H_2S conversion; (b) sulfur conversion vs. H_2S mole fraction in feed stream.

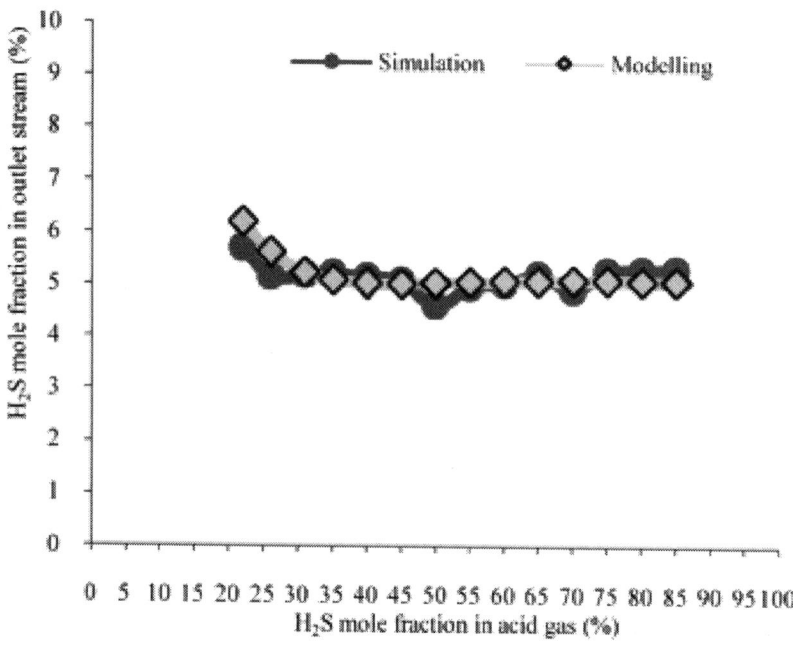

Figure 5: Estimation for H₂S content in outlet stream vs. H₂S mole fraction in feed stream.

Effect of Inlet Temperature

In this section, temperature effects have been investigated. We can predict the effect of preheating on the furnace temperature and conversion in furnace. Figure 6(a) shows the variation of furnace temperature vs. inlet temperature of acid gas. According to the figure, furnace temperature increases 4.4°C by 10°C increase in inlet temperature. Therefore, if we can increase the design temperature (52°C) to 252°C, furnace reactions will take place in 914°C.

Figure 6(b) shows the H₂S conversion by increase in the temperature. As can be seen from figure 6, H₂S conversion increases by 0.156% when inlet temperature increases 10°C. Also this figure demonstrates that H₂S conversion in furnace would increases from 72.66% to 75.74% in preheated feed (252°C).

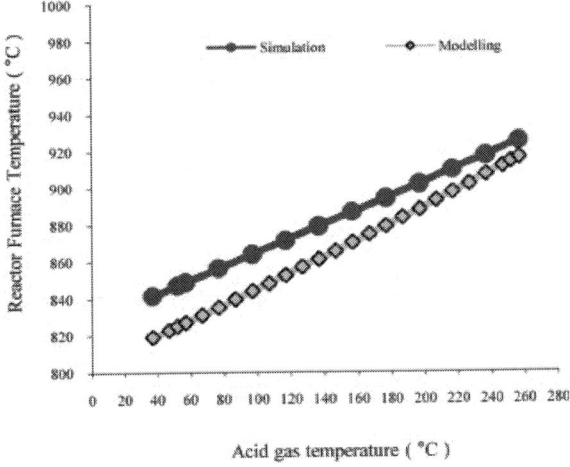

(a)

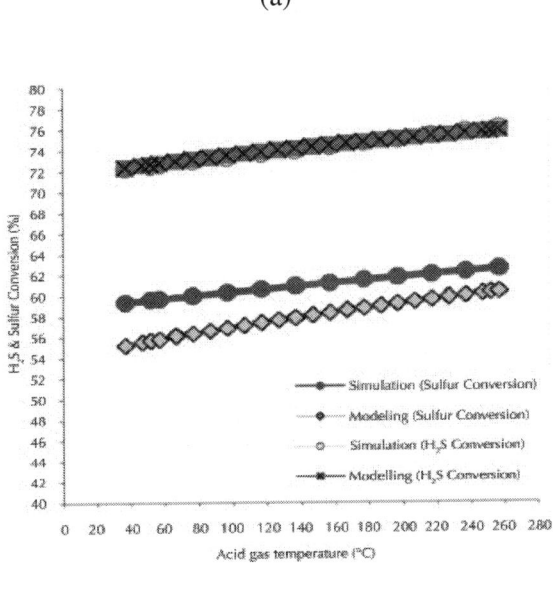

(b)

Figure 6: (a) Modified Claus furnace temperature prediction for Khangiran Gas refinery vs. inlet acid gas temperature; (b) H_2S and sulfur conversion vs. inlet acid gas temperature.

Also 10°C increase in temperature of inlet air results into 2.52°C increases in furnace temperature, 0.103% increase in H_2S conversion and 0.153% increase in sulfur conversion. If both acid gas feed and inlet air preheated separately and equally 10°C, reaction furnace temperature increases 7.1°C, H_2S conversion 0.21% and sulfur conversion 0.31%.

From a theoretical point of view, there is an optimal temperature in the furnace reactor to get the more efficient performance, maximizes sulfur production and H_2S conversion as reported in the previous sections for Claus process. A solution for this problem is fuel gas injection to the furnace in order to increase the temperature. By one percent increase fuel content in the inlet gas, furnace temperature would increase 30°C - 50°C. Calculation showed that 2000 Sm³/hr fuel injections in acid gas feed (50000 Sm³/hr) for Shahid Hasheminejad Refinery, furnace temperature would increase 130°C. More hydrocarbon content in the feed will produces more CS_2 and COS in the furnace. Increasing flow rate causes decrease in furnace capacity. It was observed the positive effect of fuel injection by increasing the temperature, led to reduction in plant capacity.

CONCLUSIONS

The reactor furnace for an industrial three-stage straight through sulfur plant with identical feed gas composition and operating conditions was molded and compared. The results of the modeling and steady state simulation have been presented in Table 5. The results showed that H_2S conversion could be promoted by an increase in hydrogen sulfide content in the feed gas. Therefore if we could enhance the H_2S concentration, sulfur conversion and overall efficiency of the furnace would improve. This also could lead to decomposition of aromatic compounds such as BTEX, additionally furnace temperature would increase. In order to increase furnace temperature, fuel injection is possible but, it must be optimized to prevent plant capacity reduction. On the other hand, reduction in CO_2 and N_2 inlet flow helps to reduce the volume of effluent and increases the furnace capacity in the Claus plant. Also results demonstrated that by utilization of heat input (preheated feed and air) in a furnace of a plant, the performance of the reactor would improve.

REFERENCES

1. M. Sassi and A. K. Gupta, "Sulfur Recovery from Acid Gas Using the Claus Process and High Temperature Air Combustion (HiTAC) Technology," American Journal of Environmental Sciences, Vol. 4, No. 5, 2008, pp. 502-511.doi:10.3844/ajessp.2008.502.511

2. K. Karan, "An Experimental and Modeling Study of Homogeneous Gas Phase Reactions Occurring in the Modified Claus Process," Ph.D. Thesis, University of Calgary, Calgary, 1998.

3. W. D. Monnery, et al., "Modeling the Modified Claus Process Reaction Furnace and the Implications on Plant Design and Recovery," The Canadian Journal of Chemical Engineering, Vol. 71, No. 5, 1993, pp. 711-724. doi:10.1002/cjce.5450710509

4. K. A. Hawboldt, "Kinetic Modeling of Key Reaction in the Modified Claus Plant Front End Furnace," Thesis, University of Calgary, Calgary, 1998.

5. S. Asadi, et al., "Effect of H_2S Concentration on the Reaction Furnace Temperature and Sulphur Recovery," International Journal of Applied Engineering Research, Vol. 1, No. 4, 2011, p. 961.

6. H. G. Paskall, "Capability of the Modified-Claus Process: A Final Report to the Department of Energy and Natural Resources of the Province of Alberta," Western Research & Development, Cheyenne, 1979.

7. H. G. Paskall, "Reaction Furnace Chemistry and Operational Models," Sulphur Recovery, Western Research, Calgary, 1981.

8. P. D. Clark, et al., "Mechanism of CS_2 Formation in the Claus Front-End Reaction Furnace," Proceedings of the 47th Annual Laurance Reid Gas Conditioning Conference, Norman, 2-5 March 1997, pp. 321-337.

9. P. D. Clark, et al., "Mechanisms of CO and COS Formation in the Claus Furnace," Industrial Engineering and Chemistry Research, Vol. 40, No. 2, 2001, pp. 497-508.doi:10.1021/ie990871l

10. P. A. Tesner, M. N. Nemirovskii and D. N. Motyl, "Kinetics of the Thermal Decomposition of Hydrogen Sulphide at 600 - 1200°C," Kinetics and Catalysis, Vol. 31, 1990, p. 1232.

11. B. Goar, et al. "Sulfur Recovery Technology," Energy Progress, Vol. 6, No. 2, 1986, pp. 71-75.

12. J. A. Sames and H. G. Paskall, "Simulation of Reaction Furnace Kinetics for Split-Flow Sulphur Plants," Sulphur Recovery, Western Research, Calgary, 1990.

13. E. A. Luinstra and P. E. Haene, "Catalyst Added to Claus Furnace Reduces Sulfur Losses," Hydrocarbon Process, Vol. 68, No. 7, 1989, pp. 53-60.

14. J. A. Sames, et al., "Evaluation of Reaction Furnace Variables in Modified Claus Plants," Sulphur Recovery, Western Research, Calgary, 1990.

Citations

CHAPTER 1

J. Parga, G. Munive, J. Valenzuela, V. Vazquez and G. Zamarripa, "Copper Recovery from Barren Cyanide Solution by Using Electrocoagulation Iron Process," Advances in Chemical Engineering and Science, Vol. 3 No. 2, 2013, pp. 150-156. doi: 10.4236/aces.2013.32018.

CHAPTER 2

M. T. Dhotre, N. G. Deen, B. Niceno, Z. Khan, and J. B. Joshi, "Large Eddy Simulation for Dispersed Bubbly Flows: A Review," International Journal of Chemical Engineering, vol. 2013, Article ID 343276, 22 pages, 2013. doi:10.1155/2013/343276.

CHAPTER 3

Manish Gupta, Jyotsna Mishra, and K. S. Pitre, "Corrosion and Inhibition Effects of Mild Steel in Hydrochloric Acid Solutions Containing Organophosphonic Acid," International Journal of Corrosion, vol. 2013, Article ID 582982, 5 pages, 2013. doi:10.1155/2013/582982.

CHAPTER 4

Vu Trieu Minh and Ahmad Majdi Abdul Rani, "Modeling and Control of Distillation Column in a Petroleum Process," Mathematical Problems in Engineering, vol. 2009, Article ID 404702, 14 pages, 2009. doi:10.1155/2009/404702.

CHAPTER 5

Mainier, F. , Fonseca, M. , Tavares, S. and Pardal, J. (2013) Quality of Electroless Ni-P (Nickel-Phosphorus) Coatings Applied in Oil Production Equipment with Salinity. Journal of Materials Science and Chemical Engineering,1, 1-8. doi: 10.4236/msce.2013.16001.

CHAPTER 6

G. Kombe, A. Temu, H. Rajabu, G. Mrema, J. Kansedo and K. Lee, "Pre-Treatment of High Free Fatty Acids Oils by Chemical Re-Esterification for Biodiesel Production—A Review," Advances in Chemical Engineering and Science, Vol. 3 No. 4, 2013, pp. 242-247. doi: 10.4236/aces.2013.34031.

CHAPTER 7

D. Campos, J. Belkouch, M. Hazi and A. Ould-Dris, "Reactivity Investigation on Iron-Titanium Oxides for a Moving Bed Chemical Looping Combustion Implementation," Advances in Chemical Engineering and Science, Vol. 3 No. 1, 2013, pp. 47-56. doi: 10.4236/aces.2013.31005.

CHAPTER 8

Jonathan W.R. Boyd, Julie Varley, The Uses of Passive Measurement of Acoustic Emissions from Chemical Engineering Processes, doi: 10.1016/S0009-2509(00)00540-6.

CHAPTER 9

R. Rezazadeh and S. Rezvantalab, "Investigation of Inlet Gas Streams Effect on the Modified Claus Reaction Furnace," Advances in Chemical Engineering and Science, Vol. 3 No. 3B, 2013, pp. 6-14. doi:10.4236/aces.2013.33A2002.

Index

A

Acoustic emission (AE) 176
Artificial neural networks (ANN)
      200

B

Bubble-induced turbulence (BIT)
      27
Bubble-Induced Turbulence (BIT)
      33

C

Chemical looping combustion
      (CLC) 149, 150
Computational fluid dynamics
      (CFD) 62
Courant-Fredrichs-Levy (CFL) 39

D

Differential pulse anodic strip-
      ping voltammetric (DPASV)
      82
Differential pulse anodic strip-
      ping voltammogram
      (DPASV) 88
Differential pulse polarogram
      (DPP) 88
Differential pulse polarography
      (DPP) 82
Direct current polarogram (DCP)
      88
Direct current polarography
      (DCP) 82
Direct Numerical Simulation
      (DNS) 23

Discrete bubble model (DBM)
    54
Discrete Fourier transform (DFT)
    183

E

Electrocoagulation (EC) 1, 6
Eulerian-Eulerian (E-E) 22
Eulerian-Lagrangian (E-L) 22

F

Fast Fourier transform (FFT) 183
Flux Corrected Transport (FCT)
    42
Free fatty acid (FFA) 135
Free fatty acids (FFA) 134

G

Genetic programming (GP) 200
Greenhouse gases (GHG) 150

H

Hydrogen cyanide (HCN) 4

L

Large eddy simulation (LES) 23,
    73
Large eddy simulations (LES) 21
Lattice-Boltzmann (LB) 59

N

Nitrilo trimethylene phosphonic
    acid (NTMP) 82
Nitrilo trimethylene phosphonic
    (NTMP) 81

O

Oxygen-carrier (OC) 150

P

Particle image velocimetry (PIV)
    58
Particle-source-in-cell (PSI-cell)
    55, 57
Probability density function (PDF)
    183
Proper orthogonal decomposition
    (POD 66

R

Re-esterification method 136
Reynolds-Averaged Navier–
    Strokes (RANS) 23
Reynolds stress model (RSM) 67
Root mean square (RMS) 182

S

Saturated calomel electrode (SCE)
    85
Silicon carbide (SiC) 115
Subgrid-scale (SGS) 24, 69

T

Thermogravimetric-thermo differ-
    ential analyzer (TGA-DTA)
    149

X

X-ray diffractometer (XRD) 155